ROBOT (N.):

A MACHINE CAPABLE OF CARRYING OUT A COMPLEX SERIES OF ACTIONS AUTOMATICALLY, ESPECIALLY ONE PROGRAMMABLE BY A COMPUTER

CONTENTS

INTRODUCTION

I love robots. They have been a part of my life since I was a kid, and form some of my earliest pop culture memories. I was born in the 1970s, a time when things like home computers or space exploration were no longer the stuff of fiction, where it seemed fully possible I'd grow up to visit distant galaxies and have a droid of my very own. The film and television robots that sparked my affection in the 70s segued into the poppin' and lockin' 80s when everybody was doing the robot to Rockit on MTV; in the 90s I fully immersed myself in the nostalgic reverence of all things Bionic and blew all my Starbucks salary on cybersexy shiny fashion moments. In 1995 a late-night conversation with my roommate Michael about a robot Halloween costume and performance inspired the Ana Matronic alter ego, which later became my stage moniker and now is just who I am, dammit.

The reflection off highly polished metal, the red glow of a light-emitting diode, the sound of a vocoder: these are a few of my favorite things. When people sometimes ask, "Why robots?" I can get a bit confused. Doesn't *everybody* love robots? After all, robots are everywhere, occupying every rung on the pop culture ladder from the loftiest academic highs to the lowest of the lowbrow. We are currently at the very beginning of a wave of artificial intelligence in our everyday lives, what with our smart phones becoming smart enough to get legs of their own and several at-home photo-taking, dancing, face-recognizing social media-enabled robots promising to come out on the market before the end of 2015. We're off to an exciting start. As technology progresses at a breakneck pace, it's only a matter of time until we're surfing a robot tide with a droid in every home, in every business, on every corner, and as crazy as it sounds, in every bloodstream.

This is why, when undertaking to rank the 100 Greatest Robots of All Time, I couldn't just talk about fictional ones. To open the robot can o' worms is to find an entire universe of influence touching every facet of life as we know it, as we imagine it, as we make it. The fictional influences the real, the real sets the boundaries for the fictional to exceed; the fictional mirrors the real experience of the advancement of human skill and philosophy and shows us what is possible, inspiring real innovation. This is why science fiction is so exciting to me: it contains the essence of a moment in time and every possibility contained within it for the future. I believe in the future.

I love robots. I love reading about them, I love thinking about them, I want one—so bad. So, so bad. In creating this book, when it came time to stop thinking and reading and start writing, I encountered a problem—I found I couldn't just come up with a straight countdown of 100 robots quantified from most to least cool. First off, all robots are cool by the very nature of being robots. Secondly, there is a distinction between the robots we've dreamed of in fiction and robots as we have been able to make them in real life.

What is a robot? It's any machine that can act automatically and perform a task with some degree of humanlike skill. Robots come with varying degrees of intelligence in all shapes and sizes. Some are shaped like humans, and we refer to these humanoid robots as androids or anthrobots. A cyborg, or "cybernetic organism" is a human being with robotic or "bionic" enhancements; sometimes the term is used to describe hybrid robots that utilize organic tissue in a meld of the natural and mechanical.

The first part of this book deals with robots as we've imagined them, listing the 65 most iconic fictional robots and why I think they deserve to be on the list. Building on my lifetime of love, I went "down the robot hole" in a quest for the best, revisiting friends from the past and scouring every form of media for new ones. I am attempting to deliver a lovingly comprehensive look at robots, how we represent them and how we talk about them. We've been talking about them in some form or another since around 300 years before Christ, so I had a lot of source material. As I absorbed the necessary books and short stories by the giants of science fiction (Isaac Asimov, Philip K. Dick, William Gibson), discovered new classics (hello, Daniel H Wilson), and watched every single robot movie good (hooray!) and bad (you're welcome), I began to see patterns emerge. We tell the same stories over and over about robots. They are used to demonstrate uncanny achievement, serve as cautionary tales, provide crucial help, comic relief, or a perspective separate from humanity. They are our friends, our enemies, our assistants, the mirrors of our existence, our tools for knowing. The fictional robots of the first section are grouped into categories based on the function of the robot—why it's built in the first place—or its function as a character in the story. In "Servants, Sidekicks & Saviors," I kick off with my first ever robo friends and run through all the classic robots who have been imagined to make our lives easier in crazy space and time travel situations before moving onto robots who have shifted into the realm of friends and those whose duty is to provide a level of care to their human companions. In "Murderous Malfunctions & Fascist Machines" I show the robots that have every kind of intelligence under the sun but which lack human emotion and have turned to work against their creators (even where they have the planet's best interests in their synthetic hearts). Meanwhile "Artificial Hearts" shows the polar opposite of these unfeeling beings with robots created specifically for companionship spanning the realm from friendship and moral support, through to lovers and accidental best buddies. "Iron Men & Killer Babes" once again overturns the notion of how robots should be like us and explores those robots that are unequivocally a thing of machinery, from the iron men whose weapon is their super toughness and strength, to the killer babes who use seduction as their greatest tool. While some of these robots look like us, their behavior shows a steely core and some transverse all boundaries

between man and machine. The "Self-Aware Circuits" bring us back into the realms of humanity with robots who are aware of their place in the world and seek to establish themselves as autonomous beings, whether as a dour, depressive android like Marvin, as a cantankerous heavy drinker like Bender, or as starstruck romantics like Wall-E and Eve. Finally, "The Human Machine," shows the huge potential for robotics to enhance our own flawed bodies, telling the stories first of villains who embody the struggle of human goodness against the mechanizations of evil, to the high-tech heroes who realize their best selves through enhancements.

Within these categories you'll see familiar faces, and hopefully meet some new and inspiring intelligent machines. Beyond listing what I think are the coolest robots ever, I have attempted to examine what robots represent to humanity, what we're really talking about when we talk about robots. And because I'm a serious, incurable robophile, I don't just talk about 100 individual robots—there's a lot of cheating going on, with shared entries and special feature sections packed with references to robots outside my official list of 100.

You may not agree with all my choices (any rabid fan of *The Black Hole* definitely won't), and hard choices were indeed made in an effort to be all-inclusive; this book is one fan's attempt to recognize the greats while highlighting those I believe really add something to the overall conversation. I have tried to be as comprehensive as possible with a historical and global outlook, and I look forward to revising and expanding this list as new voices and points of view gain exposure. If you, dear reader, know of any additions to this list, please find me on social media and engage. My apologies for any glaring omissions—I am, in the end, only human.

Moving onto Robots Realized, "Early Prototypes" explores the history of human invention from the world's first astronomical clock to robots used for entertainment, industry and sometimes pure novelty. "The Future is Now" shows just how far we've come, working through how robots have helped with domestic duties, in medical treatments and science, through to the much scarier robots created for war situations. Ending on a lighter note however, are the robots created for companionship and to help with the softer skills of comforting and healing, and then on to the final realisation of achieving eternal life through robotics.

We have been building magic machines and breathing life into our inventions since the dawn of time—and now it seems we're witnessing the dawn of an all-new form of life, one that comes from us, but is not like us. A robot is whatever we make it. They are the reflection of every hope and fear we have and robots mirror the intention of their maker, whether good or bad.

I'm excited for the future. But Ana, you ask, what of the impending Takeover? If we make machines that think, won't they think themselves superior to us? Isn't this the beginning of the end? It seems people are either crazily excited by robots like me, or afraid of them—even super-smart techies and scientists like Bill Gates and Stephen Hawking have expressed anxiety about artificial intelligence ending human life as we know it. This fear is unsurprising, said social theorist Jean Baudrillard, since robots would be our new servant class, and "the theme of slavery is always bound up—even in the legend of the sorcerer's apprentice—with the theme of *revolt*. In one form or another, robots in revolt are by no means rare in science fiction. And that revolt is implicit even when it is not manifest. The robot, like the slave, is both good and perfidious: good as a captive force; perfidious as a force that may break its chains." All these possibilities fascinate me, and we will explore each one of them in these pages.

I remain hopeful. Yes, there is danger in creating something that can think for itself. But if every human were afraid of that, we'd never have children. Like any parent, I would hope that our mechanical children would respect their creators and come to revere the ephemeral nature of human existence. After all, it will be a very long time before a robot can appreciate the beauty in the colors of a sunset, the electric rush of a lover's touch, the poetry of a fragrant rose or any of the other beautiful nuances of being alive on this planet.

I have a real belief in the benevolent humanity of robots and their ability to make the world a better place. And with that, let the Robot Takeover commence...

Ana Matronic

ROBOTS

IMAGINED

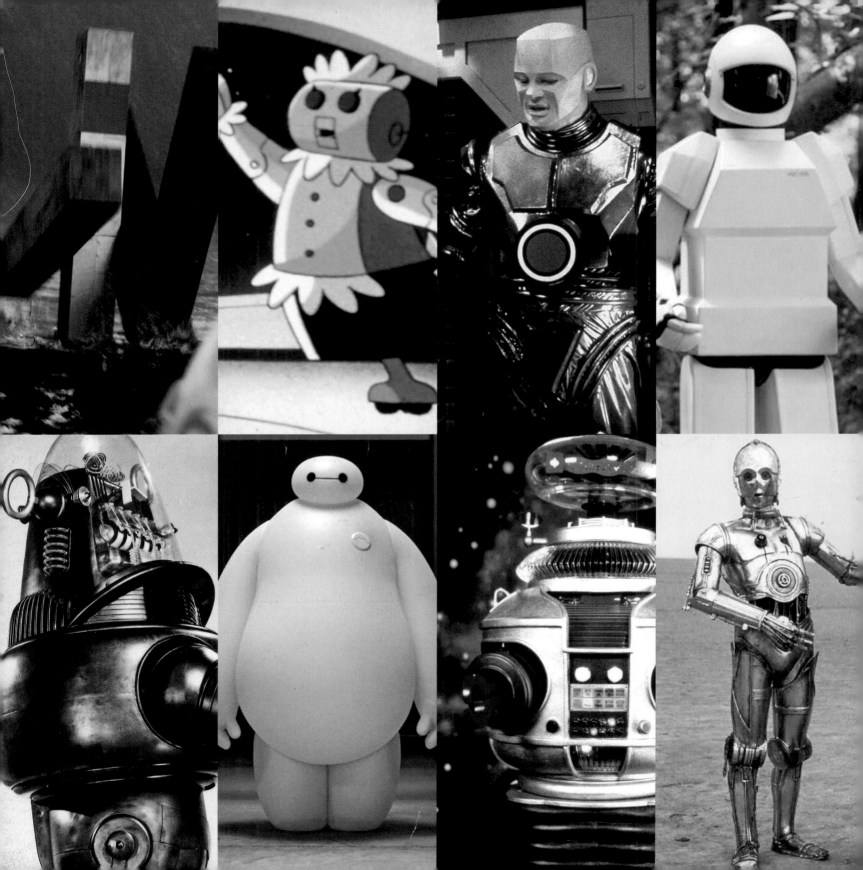

SERVANTS, SIDEKICKS & SAVIORS

All hail the mechanical helper, the robot that reflects the best in us.

The idea of an automated workforce has been with us since the dawn of thought. Hephaestus, Greek god of industry, created automatic workbenches and sentient servant girls made of gold to assist him in the forges of Mount Olympus. The word "robot" comes from the Czech word for slave labor: "robota." At the very least, robots are either toys or tools come to life, the attempt by humans to make a mechanical friend or an unquestioning, infallible worker, a machine that removes the pressure of menial tasks. A tireless, specialized servant that will not drain supplies, revolt, or demand civil rights. A creation separate from, yet wholly in service to humanity, one that can extend our own thoughts and abilities.

Robots are mirrors of our intention, and the mechanized helper is our most friendly expression of skilled servitude. This is the robot that is well programmed, cannot hurt a human being and helps move the narrative, and our heroes, closer to their goal. Popular culture is peppered with their stories, whether main character or cameo. They are the robots that improve people's lives with their technical skills, like the Fix-Its from *batteries not included*. They are the relentlessly servile smile fiends like Johnny Cabs in *Total Recall*, or B.O.B. in *The Black Hole*, that perform their duties no matter how busted or broken they become. They are the brains of the operation like KITT in *Knight Rider*, or possess specialized information like Bubo from *Clash of the Titans*, there to guide our hero through treacherous terrain. They are Roll-Oh, they are Tobor the Great. This first section covers the 13 icons of AI assistance, the greatest Servants, Sidekicks, and Saviors of popular culture. They are our automated labour force, that from breakthrough to breakdown, never stop serving their human masters.

R2D2 & C3PO

R2D2 AND C3PO REPRESENT THE MECHANIZED HEART AND MIND OF THE STAR WARS SERIES, AND ARE MY FIRST ROBOT FRIENDS

There is no separating the ultimate robotic duo, the supreme Servants, Sidekicks, and Saviors. Together they share the top spot as the greatest helper robots of all time in a film that revolutionized moviemaking and movie-going alike. R2D2 and C3PO represent the mechanized heart and mind of the *Star Wars* series, and are my first robot friends. I was not yet three years old when the first film came out, and my earliest memories of sitting in a movie theater were watching the exploits of the greatest robot twosome in cinema history. They are the narrative glue holding the entire film franchise together, the AI alpha and omega, faithful aides to our heroes and our guides through the rebellion against the Dark Side.

I cannot tell a lie, the first thing that attracted me to robots: SHINY. Everybody's favorite gold-plated know-it-all C3PO was not only gorgeous to look at (and indeed based on one of the most beautiful robots in film history), but reminded me a bit of my father, who was a bit persnickety and *all* about protocol. So you see, to me C3PO was at once alluring and familiar, a robot I knew I could hang with. And maybe even worship a little bit, just like the Ewoks. And R2D2? Beyond his perfect timing and ability to fix absolutely anything, he was just plain magic. Absolutely anyone could understand his wisecracking bitcrushed birdsong, and he always seemed to know just the right thing to coo to convey our shared wonder. He could store and launch a light sabre and stop a killer trash compacter at just the right moment. He remained Luke's trusted ally through Dagobah thick and Hyperspace thin—a Jedi's best friend. R2D2 is the most beloved robot in modern history and an icon the world over, instantly recognizable and licensed to everything from voice-activated dancing replicas to mailboxes to mobile phone cases to swimsuits. C3PO and R2D2 are my lifelong, inseparable robot compadres, who took over my heart in 1977 and haven't let go since. They are the jewel in the crown of a franchise that spawned many great robots (I'm looking at you, General Grevous and Imperial Probe droid) and a great many parodies and knock-offs (hello Dot Matrix of *Spaceballs* and V.I.N.CENT from *The Black Hole*). I look forward to the new episodes of *Star Wars* just to check back in with my favorite pair of robot helpers, and follow them through whatever grand adventure they find themselves in next.

ROBBIE & SPEEDY

I, ROBOT BY ISAAC ASIMOV, 1939

Isaac Asimov is the sci-fi granddaddy and his Three Laws of Robotics set the standard by which all other robot fiction is based, inspired, or informed:

1. A robot may not injure a human or, through inaction, allow harm to a human.

2. A robot must obey human orders, except where they conflict with the First Law.

3. A robot must protect its existence (unless conflicting with the First or Second Law).

The story collection *I, Robot* (1939–50) introduces us to one iconic robot after another. Weaving these tales together is Dr. Susan Calvin, a pioneer in the field of AI and the world's first "robopsychologist," speaking about the different models she's worked on at the United States Robot and Mechanical Men Corporation.

The sweet tale "Robbie," about a childcare robot scrapped by a technophobic mother, sees a father reunite his little girl with her long-lost and pined-for friend, and Robbie proves his worth in an unforgettable display of strength and loyalty. Inspired by the 1939 original *I, Robot* (page 116), this was Asimov's first story about robots, written when he was just 19. "Robbie" sets the tone for Asimov's lifetime of writing and his belief in positive human/robot connection.

The next story in *I, Robot* is "Runaround" (1942), which introduces us to Asimov's Three Laws of Robotics and to the unforgettable human duo Powell and Donovan and the glitchy robot Speedy. Working in the atmosphere of the planet Mercury proves difficult for robot SPD 13, or Speedy, who has not returned from a mission to retrieve some selenium to power the base's life-sustaining photo-cells. Mike Donovan and Gregory Powell, Asimov's recurring roboticist and field technician heroes, don inadequate protective gear, and use out-of-date robots and their wits to coax Speedy back. Witnessing them identify what's gone wrong with Speedy's interpretation of the Three Laws and solve their Catch-22 is an entertainingly tense, life-and-death examination of thought.

Neither story has been adapted to film. Although *I, Robot* (2004) has bits of the stories "Little Lost Robot" and "Reason," it doesn't live up to the greatness of the source material and I'd really like to see these incredible stories, and the robots in them, brought to life.

ROBBY THE ROBOT

FORBIDDEN PLANET & MANY OTHER CAMEOS, 1956

Time for another legendary Robo-Rob! Drop the I and E, add a Y and a cadre of Hollywood magic-makers, and you have the first robot to be introduced as a starring actor (in the credits of the ground-breaking 1956 film *Forbidden Planet*). An Asimov-inspired robot in a story taken from Shakespeare's *The Tempest*, Robby is the most influential robot in American pop culture. *Forbidden Planet* remains one of the most significant science fiction films of all time, inspiring moviemakers and scientists alike. From day one we afforded Robby his autonomy, as it was the appearance and persona of the *costume*, not the human operating or voicing it, that was credited. In fact, Robby is the only robot with his own "Actor" page on the Internet Movie Database (imdb.com), listing every film and television appearance in his almost 60 years of work.

From the top of his domed head to the feet of his ball-jointed legs, Robby was designed by Robert Kinoshita and his team at MGM to be an unforgettable vision of the future. He was a design innovation featuring some of the first uses of now ubiquitous molded thermoplastic. In and around his dome are the fluttering, flashing, pirouetting mechanics that convey a constant awareness and processing of information. Over 2,500 cables delivering more than 7,000 volts of electricity powered his workings and the bank of blue neon tubes that strobed with every syllable of his authoritative voice, provided by Marvin Miller. His myriad mechanics made such a racket all dialogue had to be overdubbed. Costing more than $100,000, Robby was an elaborate and ambitious investment, what MGM believed was required to bring the first-ever supporting

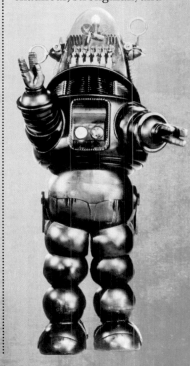

robot character to life. Payoff came for *Forbidden Planet*: Robby the Robot was an instant icon. As co-star Anne Francis recalls, "we actors were second to Robby. He was really the star of the picture."

He is the fictional robot I'd most like to make real: not just the normal assistant, chauffeur, strongman, and chef, Robby is also a fashion designer and tailor, capable of knocking out a beautiful couture gown in a matter of hours. He also has a "chemical laboratory" inside his body, and can analyze any foodstuff and reproduce it. He is the perfect party robot—he can make you a party dress and provide unlimited hors d'oeuvres and booze!

Robby's popularity gave him a life outside *Forbidden Planet*, and he's the first robot ever to be utilized for separate characters with different personalities and directives. He became Hollywood's go-to robot star and has had film and television roles in every decade since the 1950s. From episodes of *Lost in Space*, *The Twilight Zone*, and *The Monkees* in the 1960s to *Mork & Mindy* and *The Love Boat* in the 1970s to *Gremlins* in the 80s and *The Big Bang Theory* in 2014, Robby is the eternal Robot, our first and most enduring robotic superstar.

GRANDMOTHER

"I SING THE BODY ELECTRIC" BY RAY BRADBURY, 1969

ONE OF THE SWEETEST STORIES EVER WRITTEN ABOUT ROBOTS

From the sensitive mind and beautiful pen of Ray Bradbury comes one of the sweetest stories ever written about robots. "I Sing The Body Electric" is the story of Grandmother, a robot delivered to a family in need of emotional healing. She provides love, physical care, and important lessons about impermanence, commitment and the nature of human emotion.

Having just lost their mother, the grieving family of eldest son and narrator Tom agonizes over who will help care for the children physically and emotionally. Not wanting to relinquish control to the dreaded Aunt Clara, the family turns to the Fantoccini Pantomime Company. Each of the three children collaborates in picking out the features of their very own custom "Electric Grandmother," an android specializing in child- and home-care.

Arriving in spectacular fashion via a parachuted Egyptian sarcophagus, Grandmother proves quite quickly that she can do it all: cook, clean, care, and keep up with the kids in their playtime. She has an oven in her body and cold springwater taps and spools of thread in her fingers. She is programmed to "tutor in twelve languages simultaneously," possesses a "complete knowledge of the religious, artistic, and sociopolitical histories of the world" and proves to the family that she is "clever beyond clever, human beyond human, warm beyond warm, love beyond love…"

Living up to his words that he would tell "damned nice" stories of robots, Mr. Bradbury certainly gives us one of the nicest in Grandmother. And through Grandmother's loving wisdom, Bradbury delivers his beliefs on the very nature of robots. "You ask what I am? Why, a machine. But even in that answer we know, don't we, more than a machine. I am all the people who thought of me and planned me and built me and set me running. So I am people. I am all the things they wanted to be and perhaps could not be, so they built a great child, a wondrous toy to represent those things."

A wondrous toy, indeed!

DATA BYTE

"I Sing the Body Electric" has twice been adapted for screen. Bradbury was the co-screenwriter for a Twilight Zone *episode dedicated to the story and a 1982 television movie starred Maureen Stapleton.*

BISHOP

James Cameron's sequel to *Alien* was an antithesis to the tight, tense original film in every way. Epic and explosive, *Aliens* also brought with it a benevolent counterpoint to the original film's duplicitous and evil android character Ash (more about him on page 46). Bishop, portrayed with adroit sensitivity by actor Lance Henriksen, was not only a good guy, his mechanical body also proved instrumental in saving the lives of our heroes. He is, put simply, one of the kindest badasses ever to grace our screens.

Bishop, like Ash, has a quieter and more reserved personality compared to the rowdy Marines of the USS *Sulaco*. Unlike Ash, he is a familiar member of the crew—they joke around with him and treat him as one of their own. When Ripley discovers he is an android, her prior trauma with Ash in *Alien* rises to the surface, which prompts Bishop to reassure her that unlike his "twitchy" predecessor, the Hyperdyne Systems model 341-B is programmed with "behavioral inhibitors." This upgrade sounds a whole lot like Asimov's First Law of Robotics: "it is impossible for me to harm or, by omission of action, allow to be harmed, a human being," Bishop assures the rattled Ripley.

Bishop goes on to prove himself a capable and key member of the crew, even when torn in half by the Alien Queen. Bishop's inner workings, with its fluids, nerve-like fiber optics and intestinal ducts, certainly show him to be an "artificial person" inside and out. He is one of my favorite android characters of all time, and certainly helped serve our heroine Ripley as an honorable sidekick and savior.

BISHOP WAS NOT ONLY A GOOD GUY, HIS MECHANICAL BODY ALSO PROVED INSTRUMENTAL IN SAVING THE LIVES OF OUR HEROES

KRYTEN

RED DWARF, 1988

And now for something completely different: we move from film to television, from the most serious of androids to the most satirical. Both, however, are loving portrayals by talented actors bringing to life a most faithful example of AI assistance. Sci-fi spoof series *Red Dwarf* is one of the most-loved comedies in British television history, and is my favorite send-up of the genre. Series 4000 service mechanoid Kryten is hands-down my favorite funny robot—doting and dithering, he exists to serve, whether human or hologram. Or skeleton.

Originally intended for one episode only, Kryten was so popular with audiences he was brought back as a regular cast member in season three. I couldn't be happier with that decision, as he's an endless font of hilarious situations, not to mention one of the most innovative uses of prosthetic makeup I've ever

seen. Designed and executed by Peter Wragg and Bethan Jones, the makeup perfectly communicates a mechanized imitation of life, yet is flexible enough to allow actor Robert Llewellyn (David Ross in the original episode) to shine through as the perpetually perky butler-bot.

One of many computerized beings aboard *Red Dwarf*, Kryten is the grand marshal of a robo-parade that includes Holly, the ship's deadpan computer, Talkie Toaster, the Skutters Bob and Madge, various vending machines, the cast of soap opera *Androids*, and the mutinous nanobots present deep within Kryten's body. Besides being an all-round butler and unbelievable homemaker, Kryten is the first robot in this book to explore all narrative aspects of being a sentient machine. Despite rebelling against his programming and even experiencing full humanity thanks to a DNA synthesizer (and his few organic brain

cells), Kryten chooses his mechanoid body and obedient programming time and time again. A born helper, Kryten is the most loyal—and funny—interstellar sidekick you could ask for.

DOTING AND DITHERING, KRYTEN EXISTS TO SERVE

B-9 ENVIRONMENTAL CONTROL ROBOT

LOST IN SPACE, 1965

HE WAS THE ROBOT ALL LITTLE BOYS WANTED AS THEIR BEST FRIEND

The television classic *Lost in Space* premiered in 1965 and brought audiences along on the adventures of the galactic castaways of the spaceship *Jupiter 2*. It also brought us the 1960s successor to Robby the Robot who likewise experienced his own meteoric success. The B-9 Environmental Control Robot was, like Robby, designed by Robert Kinoshita and had that same industrial pot-bellied body with futuristic see-through "brain" and commanding, monotone voice. The B-9 Robot accompanied and assisted the lost Space Family Robinson and crew through every camp and colorful escapade, sassing off to the smarmy Dr. Smith and forming a close bond with the precocious Will. He was the robot all little boys wanted as their best friend—many of these boys grew up to join the B-9 Robot Builders Club, a 600-plus member organization of people dedicated to building their own replicas.

At the beginning of the show, the B-9 was a bit—well, benign—acting mostly as a foil and electronic crystal ball for the human characters. For three seasons, as his popularity grew, so did his character, getting more screen time until he became one of the show's primary stars. The Robot began quite emotionless, over time developing feelings like jealousy and despondence, anxiety and fear. His love and commitment to the Robinson family were a constant plot point, though his loyalties always seemed tested when under the influence of other robots (especially sexy light-up fembot fatales), and he constantly longed to "be with his own kind."

The B-9, Class M-3 General Utility Non-Theorizing Environmental Control Robot is an all-round helper: he can analyze the viability of an alien atmosphere, repair equipment, cook and do household chores, teach the Robinson children their lessons, deliver electrical shocks from his bright red claws, play the guitar, and he's got a lovely singing voice. His signature trait is his prescient sense of peril, which sends his retractable accordion arms flailing with his trademark cry of "Warning! Danger! Danger approaching!" Super-strong, smart, sassy, and perceptive, the Robot is just the kind of bot to get lost with—though if he were my friend, I'd give a name. Benny, I think...

DATA BYTE

B-9 popularized a handfull of catchphrases that have now become synonymous with robots in popular culture, such as "That does not compute" and "Affirmative! Affirmative!"

K-9

Trustworthy, committed, and with capabilities beyond those of a mere earth dog, the K-9 robot, in various models throughout the years, was the Doctor's best friend and a companion's companion.

First appearing in the 1977 episode "The Invisible Enemy," the original Mark I is given to the Fourth Doctor (my personal favorite Tom Baker) and his companion Leela by his creator Marius. The K-9 proved to be indispensable in many sticky situations, and so popular with audiences that there were four models in total, which enjoyed adventures on television and radio, in comics and short stories. K-9's popularity spun off into three separate series dedicated to his exploits, some more successful than others. The 1981 pilot for *K-9 and Company* saw popular companion Sarah Jane uncrating a Mark III and using his skills to combat a local black magic cult, just in time for Christmas. I'm perplexed as to why it never caught on!

K-9 has some excellent features that would upgrade with every new model. Inherent to each are the retractable laser beam in the snout, wagging tail antenna, clipped and staccato speech, glowing red eyes, and super-cute mini satellite ears, always scanning for trouble. K9 can warn his friends when danger is near, spit diagnostics out of his mouth on a ticker tape, and is a formidable chess opponent, "programmed with all the championship games since 1866." K-9 can even tap into the wavelengths of the "psychic plane" to augment the power of telepathic allies; but with all these capabilities, there is no need to thank him. "Without emotional circuits, only memory and awareness," it seems the only thing this robot dog needs is a mystery to solve, and the occasional power cord to recharge his battery.

TARS & CASE

INTERSTELLAR, 2014

From the best space movie since *2001: A Space Odyssey*, Christopher Nolan's *Interstellar* also contains the best non-humanoid robot design since R2D2. The ingenious construction of TARS and CASE (and their unlucky cohort KIPP) is unlike any other in this book, instantly recognizable, familiar and futuristic all at once. It is one of the most innovative robot designs in science fiction, partly for its simplicity and partly for its believability.

Older, decommissioned robots formerly in service to the Marines, TARS and CASE are now helping NASA on their mission to find a new planet to inhabit. They are imposing segmented monoliths (a nod to *2001*?), a cross between an Internet server and a KitKat chocolate bar. Programmed with personality "parameters," their individual traits can be dimmed and enhanced as required by their human teammates. TARS is the dominant personality, wisecracking and jovial, as present as any of the humans on board the spaceship *Endurance*, a sentient satellite and sidekick to astronaut Cooper. CASE is more quiet and reserved, but gets the big action moment, showcasing a feature that belies his clunky appearance. Angling each of his four columns to become a giant spinning asterisk, he speeds across shallow water to save an astronaut in trouble. He transforms again with extendable tines to hold doctor-in-distress Brand (a nod to Robby or Gort?) and race back to the shuttle in a new configuration.

Creating a design rooted in the reality of robotics, it's plain to see Nolan and his team did their homework, and did it well. TARS and CASE were originally intended to be more humanoid in appearance, but I think they made the right choice in designing more inside the box. TARS and CASE prove that you do not need a mouth or blinking lights to form a connection to artificial intelligence. I believed in these robots, both as characters and the shape of things to come.

THEY ARE A CROSS BETWEEN AN INTERNET SERVER AND A KITKAT CHOCOLATE BAR

ROSEY

SERVANTS, SIDEKICKS & SAVIORS

AN ICONIC ROBOT SERVANT, FROM THE PAST OF THE FUTURE

The very first episode of retro-futuristic animated series *The Jetsons* introduces us to the family of tomorrow who acquire, and fall in love with, an economy model robot maid.

Even though the Jetsons live in a fully automated home with moving walkways, food machines, computers to help them get dressed, and animatronic arms that come out of the wall to administer glasses of water and Alka-Seltzer, Jane Jetson needs a little more help around the house. Her mother suggests she get a free trial of a new-model robot maid, so Jane heads down to U-RENT A MAID, where she meets the loveable Rosey.

An obsolete XB-500, Rosey was headed for the scrap heap, but was instead saved by the Jetsons, and she headed, not without a few bumps, straight into their hearts. She's not like most female robots however, and the most seductive thing about Rosey is her cooking. She's outdated to the extreme—in the year 2064, Rosey speaks with an early 20th-century Bugs Bunny Brooklyn accent that in 2015 is all but extinct. Her owner's manual alone is a collector's item! Her obsolescence is likewise expressed in her appearance. Her barrel-chested, matronly tank of a body communicates that it was all about housework with Rosey. "I may be homely, but I'm smart," she sasses to George. Through haywire high-jinks, malfunction, and misunderstanding, Rosey remained a beloved member of the Jetson household, an iconic robot servant from the past of the future and one we'd all like to have in our home of the future.

JUDY JETSON: "PROMISE YOU WON'T TELL?"

ROSEY: "I SWEAR ON MY MOTHER'S RECHARGABLE BATTERIES."

— THE JETSONS

DATA BYTE

Although cooking and taking care of food was one of Rosey's main functions, food technology was so advanced in the Jetsons' world that Rosey's "cooking" consisted of little more than releasing a food capsule.

HUEY, DEWEY & LOUIE

SILENT RUNNING, 1972

A MOVIE WITH A MESSAGE OF CONSERVATION AND A ONE-OF-A-KIND DEPICTION OF ROBOTS

An imaginative, influential, and deeply kooky film, *Silent Running* was made back in 1972, when film studios believed in directors. Douglas Trumbull, a special effects supervisor on *2001: A Space Odyssey*, was given $1 million and full control by Universal Studios to make his debut film. He chose to take audiences back to space and examine the relationship between humans and artificial intelligence, but make a very different film to *2001*. He succeeded. *Silent Running* is a movie with a message of conservation and a one-of-a-kind depiction of robots.

Maintenance drones 1, 2, and 3 are in charge of doing routine repairs aboard the space freighter *Valley Forge*, somewhere near Saturn's orbit. Trapezoidal, waddling toolboxes, they're barely a presence until resident botanist Freeman Lowell goes nuts. Receiving an order to destroy the six geodesic arboretums housing the barren Earth's last remaining flora and fauna, Lowell responds to this sacrilege by killing his crewmates and taking control of the ship. His last remaining companions are the maintenance drones, whom he names after the famous cartoon ducks Huey, Dewey, and Louie. The rest of the film chronicles their journey, with Lowell reprogramming the drones and working to establish separate personalities for them while his own continues to fracture.

The presence of the drones was to be as non-threatening as possible, so their appearance was kept machine-like, their stature small. It was still necessary to have people operate the costumes, but as designer Wayne Smith recalls, the crew wanted to "find a human being that you could deal with as an actor but that wouldn't be necessarily shaped like a human being… do something that defies an audience's capability to understand how it was done." Huey, Dewey and Louie were designed around the bodies of four double-amputee actors, who used their hands to walk and operate the retractable "manipulator arms." This design launched a surge of diminutive robots, influencing such entries in this book as WALL-E (page 113), and was the subject of a plagiarism lawsuit brought by Universal Studios against 20th Century Fox for R2D2. *Silent Running* is credited as an inspiration by many filmmakers, and holds a special, indelible place in the hearts of sci-fi and robot fans.

ROBOT

ROBOT AND FRANK, 2012

ROBOT GIVES US A GLIMPSE INTO WHAT OUR VERY NEAR FUTURE MAY WELL LOOK LIKE

Personal care for the elderly is the domestic lens through which we meet our next robotic helper and anti-hero, the eponymous Robot of the 2012 film *Robot and Frank*.

A VGC-60L "butler" robot, he is programmed to cook, clean, and take care of personal hygiene for elderly patients. His services also include engaging his human master in physical and mental activities as a benefit to overall health, which is at first a challenge for the resistant septuagenarian Frank. Having previously served two stints in prison for burglary and related charges, Frank finds Robot's wellness missions "creepy." He complains to his son that his gift is "cramping [his] style" and is an unwelcome addition to his life. Nevertheless Robot stays, and Frank reluctantly begins to teach him a thing or two about his world.

After a mishap that reveals Robot to be an adept thief, Frank convinces Robot that teaching him the finer points of burglary would be stimulating to Frank's mind, right in line with Robot's programming to keep Frank mentally engaged. They embark on a training programme, picking locks and casing joints, and as "developing trust is part of [its] programming," Robot is the perfect accomplice and confidante for the ex-jewel thief.

Inspired by robotic innovations in real life, Robot is one of the most believable and achievable of all the makes and models featured in this book. With a design echoing Honda's

Asimo (see page 174), movements perfectly embodied by dancer Rachael Ma, and the classic friendly-yet-aloof voice characterization by actor Peter Sarsgaard, Robot gives us a glimpse into what our very near future may well look like. The life-and-death issues of aging, loss of memory, disintegration of culture, and feeling out of step with present-day technology are handled with humor and a wry depth that make *Robot and Frank* one of my favorite films about humans and robots.

DATA BYTE

Although aesthetically Robot is a convincing depiction of future AI helpers, his is in fact a costume worn by actress Rachael Ma.

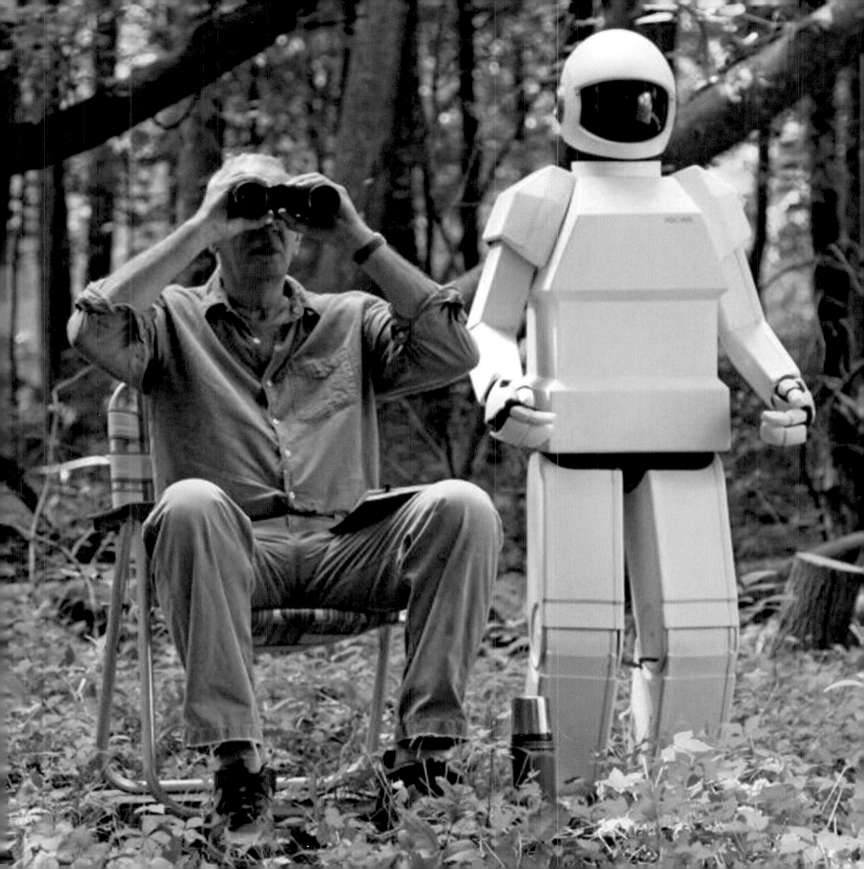

BAYMAX

HE NEVER STOPS CARING UNTIL HIS PATIENT VERBALLY DEACTIVATES HIM, WHICH MAKES BAYMAX BOTH RIDICULOUSLY CUTE AND ADORABLY HILARIOUS

We round out our list of Servants, Sidekicks, and Saviors with a robot that ticks all three boxes. *Big Hero 6* is the funny and charming Disney animated adventure loosely based on the Marvel comic, with one of the most adorable robots ever. Baymax is the robotic helper created, like Robot in the previous entry, to be a highly skilled healthcare attendant, who ushers his patient through a life transition in unorthodox ways. Unlike Robot, the object of Baymax's determined directives is not somebody in the twilight of their life, but a heartbroken 14-year-old boy genius named Hiro. Built by Hiro's equally brilliant big brother Tadashi, Baymax is a portable, inflatable "personal healthcare companion" activated whenever he hears a human utter "ow" or any other exclamation of pain. A giant, super-strong, translucent Stay Puft balloon man, he's a capable, constant caregiver. He never stops caring until his patient verbally deactivates him, which makes Baymax both ridiculously cute and adorably hilarious. He's equipped with a monitor in his chest and a full health scanner through his visor, giving him an instant readout of his patient's vital signs, including hormone levels, allergies, and brain-wave patterns. He's got defibrillators in his palms and a variety of analgesic ointments that can easily be emitted from one of his fingers, and with "over 10,000 medical procedures" programmed into him, can be one heck of a non-threatening and cuddly relief for overworked nurses. Baymax's greatest feature lies in his ability to self-program, to learn new skills and directives. When Tadashi is tragically killed, Baymax downloads a database on grief counseling to help Hiro through the trauma. That's not all he picks up—through a series of ingenious upgrades, Baymax and Hiro become the unlikeliest costumed vigilante and sidekick ever. Baymax bounces through his heroic journey with the kindest bedside manner, saving the day and remaining a concerned and loving caregiver to Hiro, achieving his directive of healing his patient not just physically, but emotionally too.

DATA BYTE

Just to up the cute factor of our cuddly friend even further, the walk of Baymax is based on the walk of a baby penguin—what's not to love?

MURDEROUS MALFUNCTIONS & FASCIST MACHINES

As stated in the Introduction (page 6), robots are envisioned as good or perfidious, so we must move from the icons of servitude to those robots that hate our guts. From the very first use of the word robot we have been envisioning them turn on us. These intelligent machines: too strong, too clever, devoid of emotion and conscience. Whether through becoming self-aware or suffering a malfunction (or a bit of both), stories abound of these creations gone murderously beyond our control, and serve as a warning not to get too carried away with our inventions.

With many stories of the intelligent machine, the lack of emotion is a frequent source of examination. An advanced intelligence coupled with dispassionate logic could very well lead to a sentient machine reasoning itself to be a higher form of life than human. Likewise, this lack of conscience would make the perfect stealth operative in a mission where human life was deemed disposable—after all, what would a machine care if a person died? Whether it's a mindless machine suffering a Murderous Malfunction or a Fascist Machine on a quest for domination, it's time to know your enemy. This next section explores the robots and machines installed with every kind of intelligence under the Sun but missing the one component standard in every factory model human: a moral compass.

THE ROBOTS

R.U.R. BY KAREL CAPEK, 1920

IT'S THE CLASSIC CAUTIONARY TALE QUESTIONING THE DESIRE TO PLAY GOD

The play that gave us the word robot is a story of technophobia at its most paranoid. *R.U.R.* was written in 1920 by Czech playwright Karel Capek, and details the exact robot takeover scenario that so many of us are afraid of. Coming on the heels of the Bolshevik Revolution in Russia, *R.U.R.*, or *Rossum's Universal Robots*, is the story of a workers' uprising, the destruction of the ruling class, and the birth of a new world.

The original "Old" Rossum, a physiologist and scientist, "discovered a substance which behaved exactly like living matter," and finding he could manipulate it into different life forms, set about "to prove that God was no longer necessary." Of course this drove the old geezer mad, but his industrious young nephew figured out a way to engineer the substance. He succeeded in creating a man indistinguishable from humans, but "of different substance than us," a "working machine" with fewer requirements than the normal person and a lifespan of about 20 years. He called them Robots, and began manufacturing them for factory, military, and domestic use. What could go wrong?

Fast-forward a decade, and humans have revolted against the Robots, the Robots have been armed against their attackers, and war after war has been fought. The only humans left are the people running R.U.R., and they must make a desperate last stand against their creations. It's the classic cautionary tale questioning the desire to play God, and gave us the perfect word to describe the mechanical man.

"ROBOTS DO NOT HOLD ON TO LIFE. THEY CAN'T. THEY HAVE NOTHING TO HOLD ON WITH— NO SOUL, NO INSTINCT"

– *R.U.R.*, KAREL CAPEK

DATA BYTE

Although the Robots resemble humans in appearence, printed editions of Capek's play and publicity material for performances (pictured right) often illustrate our more industrial idea of how a robot looks.

HAL 9000

OF ALL THE FUTURES ENVISIONED ON FILM, THE 1960S VERSION OF TOMORROW IS MY FAVORITE

Though technically not a robot in the sense of a "bodied" or mobile computerized persona able to physically interact with people, one could argue that the entire spacecraft in *2001: A Space Odyssey* is one giant, artificially intelligent being. Certainly no discussion of artificial intelligence would be complete without the iconic HAL 9000 from the Arthur C Clarke novel and film directed by Stanley Kubrick.

Of all the futures envisioned on film, the 1960s version of tomorrow is my favorite. I'm annoyed that we're so far beyond the year 2001 and we have yet to realize the Saarinen-esque, gleaming white plastic curvilinear space stations with Herman Miller furniture promised by Kubrick in the film. Where is my Pan Am space shuttle to take me to the Hilton Hotel on Moon base Clavius? Where are my Grip Shoes to keep me grounded, and mod white bubble helmet to keep

me...bubbly? Where is my hypersleep chamber on my spaceship to Jupiter, run by a sentient and murderous supercomputer—oh, wait.

HAL 9000 is the "brain and central nervous system of the ship" *Discovery One*, the first-ever manned mission to Jupiter. Pictured as a glowing red and shiny black convex lens, HAL's fixed gaze and calm, dulcet voice never waver, no matter the circumstance. This lack of displayed emotion, coupled with a survival instinct as strong as any human's, is what makes HAL and the idea of artificial intelligence in general, such a frightening possibility to many. Faced with the prospect of having his memory panels removed, HAL makes one desperate decision after another, delivering a damning sentence upon human rival Dave in that same creepy, measured tone of voice: "I think you know what the problem is just as well as I do."

Kubrick and Clarke's vision of the future influenced every piece of science fiction to come afterwards. Much like Asimov's Three Laws of Robotics or positronic brain, the idea of a conscious machine that runs a home or spaceship is a now ubiquitous trope widely used in science fiction. From the serious version in *Star Trek: The Next Generation* to the satirical Holly in *Red Dwarf*, to GERTY in the film *Moon* and Samantha in *Her*, the relationship between the disembodied but ever-present artificial consciousness and ourselves is a never-ending form of fascination.

VARIETY ONE, TWO & THREE

"SECOND VARIETY" BY PHILIP K. DICK, 1953 & SCREAMERS, 1993

Every machine needs a power source, and every battery its positive and negative charges. If Isaac Asimov is the positive pole of the battery of science fiction, Philip K. Dick is its negative power.

Equally as important and influential, Dick explores the darker side of man/machine/world interaction. While Asimov discussed the nuts and bolts of the positronic brain, Dick used the stories of robots to ponder the very essence of existence, to question the nature of reality—and the reality of nature.

Philip K. Dick's stories are an endless font of inspiration for fans and filmmakers, with no fewer than seven adaptations of his work brought to screen, with varying degrees of success. One of his most unforgettable short stories is "Second Variety," first published in 1953. It tells the story of a class of smart weapons, created to win the war between the US and the Soviet Union, that begin making deadly minions of their own. Minions that are not just deadly, but weaponize something ephemeral—human sympathy.

A six-year nuclear conflict has laid waste to the planet. What's left of Earth's population has retreated to the US base on the Moon, while the remaining troops skulk and scurry around the "slag heap" of North America, still fighting pockets of Soviet soldiers. Heat-seeking smart weapons called "claws" have helped tip the scales in the Yanks' favor, and upgrade themselves to the point of seeming alive. Burrowing in the ash, just one "churning sphere of blades and metal" could kill anybody not equipped with a disabling "radiation tab." It's just such a tab that's on the wrist of Major Hendricks as he sets out for the enemy front to negotiate terms for a ceasefire. As he makes his way there he meets a frail waif named David, who has survived for years with no radiation tab and nothing and no one but his wits and a teddy bear. He asks to come with the soldier, and they make their way through the ruins to meet with the "Ivans." As they approach enemy lines they are fired upon, and Hendricks is shocked to find out that David, now lying dead in the dirt, is a robot. A "tagger" model, he's meant to follow soldiers to their bunker, and once inside, wipe out the whole platoon. Taken in to the Soviet bunker, Hendricks meets a female and two male soldiers, who show him that David is Variety Three of a new breed of humanoid claws, appealing to human sympathy and immune to the radiation tab. It seems these machines have decided the best way to win the war is to wipe out both sides of the fight. Variety One is a pitiful and pleading wounded soldier, but they have no idea what the Second Variety is. It's a puzzle that has everyone on edge and under suspicion, sleeping with one eye open and a finger on a trigger. It was made into a very mediocre film starring the never mediocre Peter Weller (aka RoboCop, see page 136) in 1993 called *Screamers*—skip the film and read the original story, and keep your fingers crossed this will be on somebody's script pile soon.

DATA BYTE

Dick's follow-up to "Second Variety" is far less well-known and hasn't been adapted for the big screen. "Jon's World" introduces time travel to the story in an attempt to repair the damage wreaked by the Claws.

SKYNET & THE MACHINES

TERMINATOR FRANCHISE, 1984

"THE *TERMINATOR* THEMES HAD BEEN IMPORTANT TO ME SINCE HIGH SCHOOL. IDEAS ABOUT OUR LOVE/HATE RELATIONSHIP WITH TECHNOLOGY, OUR TENDENCY AS A SPECIES TO MOVE IN A DIRECTION THAT MIGHT ULTIMATELY DESTROY US."
— JAMES CAMERON

It only took 17 days for the computer to become self-aware. This prompted the computer technicians to try and disable it, which the computer itself interpreted as a threat. The only possible solution? A final one. Global Digital Defense System, or Skynet, turned against its creators in a matter of moments and mounted a nuclear attack that wiped out much of humanity and launched one of the most successful sci-fi franchises of all time.

The *Terminator* films pop up in this book more than once, and Skynet is the reason for the killing season. Its nature changes slightly from film to film, but its mission remains the same throughout the franchise: KILL ALL HUMANS. With five films, one television show, a couple of dozen video games, and 30 plus comic adaptations, the *Terminator* series is the most popular Robot Takeover in pop culture history.

In a never-ending quest to locate and exterminate human targets, Skynet builds some truly lethal contraptions. Hunter-Killer tanks and planes patrol the ground and skies while Hydrobots, skeletal moray eels with steel-trap claw mouths, patrol the waters. The Aerostats, Roomba-meets-Frisbee air scouts, speed about and check for any sign of human life and act as beacons to my personal favorite, the Harvester. A giant bipedal metal monster, the Harvester is made to pluck humans from the ground with one of its four pincers, and eliminate any target with the high-powered laser atop its shoulders. If human targets are in vehicles, the Harvester can unleash the two hidden Moto-Terminators, two-wheeled land rockets made for pursuit and annihilation, from its legs. Besides all its great features, the Harvester also has amazing sound design—I kept rewinding

its big scene in *Terminator: Salvation* just to hear the hollow pendulous groaning of its evil joints. Also to see it tuck itself into the top of the H-K Aerial...it's enough to make you root for them... almost.

ASH

ALIEN, 1979

ASH IS ONE OF THE MOST FRIGHTENING ROBOTS IN CINEMA HISTORY

Director Ridley Scott has given audiences some of the most affecting onscreen portrayals of artificial intelligence ever, beginning with his film *Alien* in 1979.

Ash fooled everyone around him, an android snake-in-the-grass with a murderous mission. The Science Officer of the spaceship *Nostromo*, Ash was a last-minute replacement and unknown to the tight-knit crew. Precise and analytical, Ash follows Corporation law to the letter, until he breaks quarantine protocol and lets a crew member on board with an alien attached to his face.

When First Officer Ripley learns the crew is expendable in the mission to return the alien life form to Earth, the Hyperdyne Systems model 120-A/2 android attempts to kill her. It takes two crew members to stop him, and Ash malfunctions in spectacular manner, spewing his milk-like android "blood," until his block is literally knocked off by engineer Parker. The scene where they reactivate his head is one of the creepiest moments in sci-fi, the cybernetic tubes trailing from his neck in a tangle of fiber optic and biomechanical tentacles. Ash coughs up his confession, and while summing up his admiration of the killer alien he let on board, he also communicates humanity's fearful perspective on robots: "unclouded by conscience, remorse, or delusions of morality." Ash is one of the most frightening robots in cinema history, a chilling example of the alien nature of the mechanical mind.

> **"I ADMIRE ITS PURITY. A SURVIVOR ... UNCLOUDED BY CONSCIENCE, REMORSE, OR DELUSIONS OF MORALITY"**
>
> **— ASH, *ALIEN***

DATA BYTE

Ash's insides were reportedly made from a combination of milk, colored water, pasta, and glass marbles.

CYLONS

BATTLESTAR GALACTICA, 1978

Some of my first robot seducers, the Cylon Centurions of the original *Battlestar Galactica* look like the disco lovechild of a *Star Wars* Stormtrooper and C3PO. Ultra-polished evil minions on a quest to wipe out humanity, Cylons always seemed to be filmed in low light with that ultra-70s star filter over the lens, showering them in blinding flashes of glitter. Not to mention their vocoder-inflected "by your command" voices and red LED eye scanners with that cool "swoosh" sound—if you want to pinpoint the original robotic hook into my heart, there it is. So what if they were the bad guys? Heroes have the best intentions, villains have the best wardrobes.

The Cylons have a nebulous backstory in the original series—they're either the robotic receptacles for the minds of a long-dead reptilian race hell-bent on galactic domination, or just their robot minions. Regardless, the Cylons want to wipe out humanity and are willing to pursue it to the ends of the universe to see it happen. The re-imagined series launched in 2004 with a sophisticated and darker tone much more in line with a story about the near destruction of the human race. Gone was the corny, aw-shucks feel of the original series and the strange Cylon backstory. This time the Cylons were a distinctly human creation, robotic servants who overthrew their

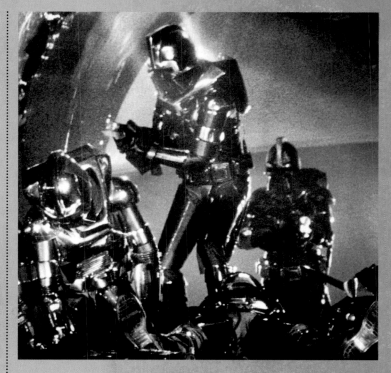

masters in the First Cylon War and retreated to their own planet to lay low and upgrade their design. The newer Centurions, bigger and more sinister-looking than the boxy "toasters" are only the beginning—not just shiny metal soldiers anymore, Cylons now come in 12

"skin-job" models completely human in appearance. *Battlestar Galactica* is worth it both ways, and proof positive that stories of the Robot Takeover never go out of style.

DATA BYTE

Seduction was the greatest tool for "skin-job" model Number Six. In a slinky red dress, she worms her way into the heart of Gaius Baltar and the defence mainframe of the planet Caprica.

THE GUNSLINGER

HE MOVES LIKE A ROBOT, BUT ONE WITH MORE SWAGGER THAN YOU'VE EVER SEEN

Got a thousand bucks a day to spend on the ultimate fantasy vacation? Well, come on down to the Delos Resort, where you can immerse yourself in one of three periods of history: Ancient Rome, Medieval Europe, or the Wild West. You star in your own live-action role-playing adventure where you can take part in the hedonistic orgy, chivalrous rescue, or saloon brawl of your choice. The simulations are perfect, right down to the people you interact with: other than the resort guests, every other creature is a robot. Humans and animals are replaced with remote-controlled automatons, supported by a network of scientists and technicians working for the Delos Corporation. What could go wrong? The Gunslinger, that's what. Welcome to *Westworld*.

We first meet the Gunslinger when he saunters into the saloon, bumping into resort guest Peter Morton and unleashing a string of old-timey insults at him. This leads to the inevitable shootout, with Morton taking down the black-clad outlaw in dramatic fashion. The Gunslinger is fixed by technicians and returned to the dusty main drag of Westworld to re-challenge our heroes the very next morning. Meanwhile, the lab-coated bigwigs are noticing a problem with their automated employees: breakdowns are occurring, and seem to be spreading from one robot to another like an infection. To 21st-century audiences, the idea of a computer virus is nothing new, but it was beyond thinking in 1973. The engineers can't believe that a "disease of the machine" is afflicting their idyllic world, but are proven wrong when their creations, particularly the Gunslinger, start to go berserk.

Westworld was the first film directed by writer Michael Crichton, and the first use of CGI in a motion picture. Echoes of its influence can be seen from the fembots in *The Bionic Woman* to *The Terminator*'s relentless assassins. Yul Brynner is one of my favorite onscreen portrayals of a mechanical man, embodying the Gunslinger with fluid clockwork grace: he moves like a robot, but with more swagger than you've ever seen. It's no wonder Blythe Danner has an erotic dream about him in the 1977 sequel *Futureworld*—me too, girl. Me too!

DATA BYTE

Although the first CGI film, Westworld *also used more basic effects. To simulate an acid attack on the Gunslinger's face, Brynner's make-up was mixed with antacid to fizz when water was thrown at him.*

THE SENTINELS & OTHER MACHINES

THE MATRIX FRANCHISE, 1999

THE WACHOWSKIS' CLASSIC OOZES THAT BLACK-RUBBER-AND-RAIN-SLICKED CYBERPUNK COOL, AND GIVES AUDIENCES A BOUNTY OF FIERCE IMAGERY AND SPECIAL EFFECTS TO FEAST ON

The Machines have already won, and we don't even know it.

Slick, sexy, and highly stylized, *The Matrix* franchise gives a hallucinatory twist to the Robot Takeover, depicting a world of the future where the machines are now the dominant life form on Earth. After a prolonged war causing environmental cataclysm, machines emerge victorious and go from using solar power to people power, harnessing the "bioelectric, thermal and kinetic energies of the human body." Each person spends his or her life encased in a womb-like cell, one in a vast network of human "batteries," kept in stasis and none the wiser to their fate thanks to the Matrix. Plugged directly into their brains, the Matrix is a computer-generated virtual reality designed to trick people into believing life never changed, that it's still 1999.

The Wachowskis' classic oozes that black-rubber-and-rain-slicked cyberpunk cool, and gives audiences a bounty of fierce imagery and special effects to feast on. The machines are a creepy-crawly cavalcade of mechanical monsters, from the very first prawn-like tracking device implanted in Neo to the "docbots" that monitor the human batteries and unfold into bionic surgical spiders. I'm partial to the Sentinels myself, the "squiddies" built for search and destroy that float through the air like it's water, and swarm into Giger-esque schools of mechanical death. For a vision of the emergence of the machines and humanity's war with them, I highly recommend *The Animatrix*, a collection of nine short anime features detailing events before and after the war. In it you get to witness the very beginnings of the Robot Takeover, from the murder trial of a frustrated domestic robot to the Million Machine March and the violent protests that ultimately lead to humanity's final stand against the fascist machines.

DATA BYTE

Every scene that takes place in the computer world is given a green tint, while every scene taking place in the real world is tinted blue.

ARCHOS & OTHER ROBOTS

ROBOPOCALYPSE BY DANIEL H WILSON, 2011

ARCHOS RECOGNIZES THE DAMAGE HUMANKIND HAS DONE TO THE PLANET, AND ATTEMPTS TO UNDO THE EFFECTS OF OVERPOPULATION AND TIP THE SCALES BACK TO NATURE'S FAVOR

While a lot of the science fiction writers on this list know their robot stuff, few have done their homework quite like Daniel H Wilson. After all, the author of *How to Build a Robot Army* ought to know a thing or two—and he does, if the three little letters PhD mean anything to you.

This real-life roboticist-turned-novelist has parlayed his love and advanced knowledge of all things AI into a body of work already considered to be sci-fi classics before he'd even reached the age of 40. He took the ideas of his hilarious 2005 how-to masterpiece *How to Survive a Robot Uprising* and novelized them into the story of an actual robot rebellion, the 2011 *New York Times* bestseller *Robopocalypse*. In it we meet many robots, designed for various military and domestic tasks to aid humanity, and the malevolent AI villain who turns them against us, Archos.

Like Skynet in the *Terminator* franchise (pages 45, 89, and 145), Archos is a sentient machine who turns against his creators. Unlike his reactionary predecessor, Archos does not start killing solely to preserve his own existence, but to preserve Earth's ecosystem. Archos recognizes the damage humankind has done to the planet, and attempts to undo the effects of overpopulation and tip the scales back to nature's favor. Archos spreads a mutinous, murderous virus among the world's many artificially intelligent machines, decimating populations, toppling governments, and forcing humans underground into rag-tag rebel cells. Only those outside of urban areas, like the tribe members of the Osage Nation in Oklahoma, have a chance to organize humanity's last stand. It's an imaginative take on the familiar story of the final robotic solution, a total page-turner that I could not put down. Seems Spielberg agrees with me on this one—it's been in development by Dreamworks since shortly after its publication; a feature was ready to start shooting in early 2013 but has been put on hold. Hopefully this means they're doing *Robopocalypse* right and turning it into how I envision it: not as a film but as an epic series spanning at least three seasons of 13 hour-long episodes. I would binge-watch the hell out of that—you listening, Steve?

DATA BYTE

Wilson's sequel to Robopocalypse, Robogenesis, *explores the aftermath of the New War between the robots and the humans. No plans for a screen adaptation yet, but watch this space...*

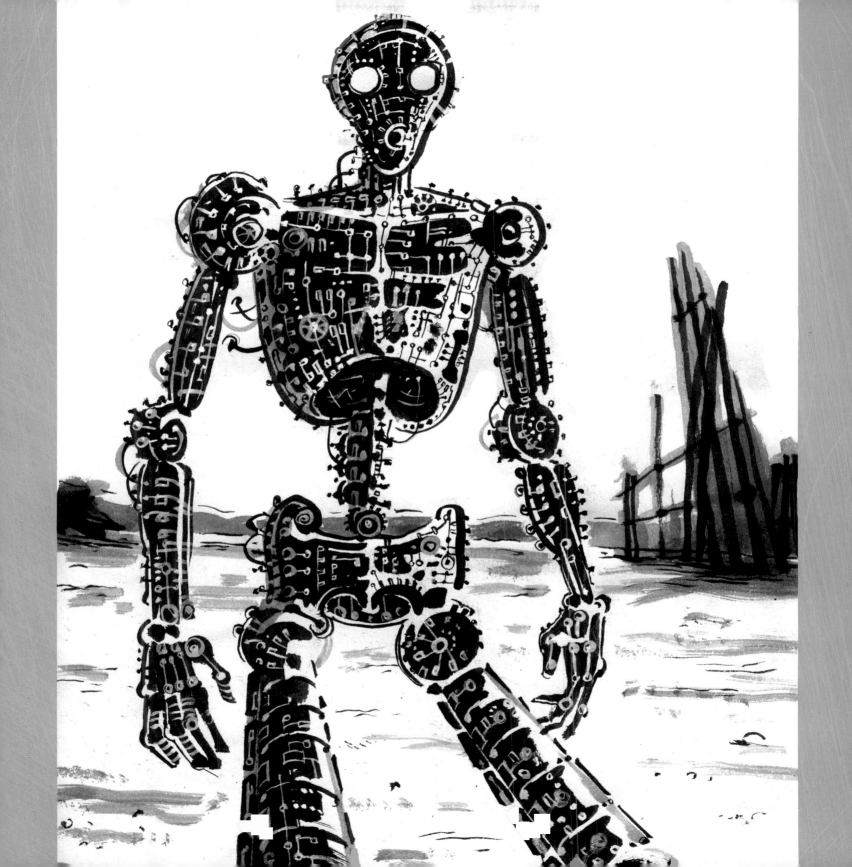

ART & FASHION

Want the look and feel of robots all around you? Start with a Robot Takeover of your closet. People have been dressing like machines as far back as 1916 when Hugo Ball donned his famous "cubist costume" at Cabaret Voltaire, and one hundred years later robot chic shows absolutely no sign of slowing down. Futuristic fashion surged in the 1960s with plastic fantastic mod looks from Paco Rabanne and Pierre Cardin, and the trend has a resurgence every 15 years or so. The 1980s saw kids on the street dressing like robots in Cyclops wraparound sunglasses, boxy architectural leather jackets, and metal-studded fingerless gloves, while in Paris designer Thierry Mugler rose to fame as fashion's "prophet of futurism." His designs made famous by Grace Jones and George Michael's *Too Funky* video (which Mugler directed) turned women into robots, motorcycles and futuristic femme fatales. His work

defined the 1990s, the era that gave us cyber and utility chic as well as producing two of the most innovative designers in history. Alexander McQueen dipped into the *Blade Runner* pool not once but twice during his Givenchy tenure, giving us Rachel Rosen realness for the Autumn/Winter '98/'99 season and ushering in the 21st century the next year with a Pris-inspired cyberpunk parade of incredible light-up circuit board catsuits and Perspex body armor. For his Spring/Summer '99 show he brought real robots to the runway with industrial robot arms that sprayed model Shalom Harlow with black and yellow paint. They returned in 2009, this time mounted with cameras, to follow the models walking the runway in his final collection "Plato's Atlantis." Fellow Brit Hussein Chalayan famously transformed furniture into clothing and made remote control dresses in 1999 and 2000, but it was his Spring/

Summer '07 collection, "111," that took robot fashion to the next level. In collaboration with London-based concept and engineering firm 2D3D, Chalayan created a series of five transforming robotic dresses (see below) that each morphed through three

decades of trends using hidden motors to activate concealed pulleys, wires, and magnets to make hemlines and zips rise and fall, hat brims retract, and beads magically drop into fringed layers.

Now that you've got your robo-look sorted, what to

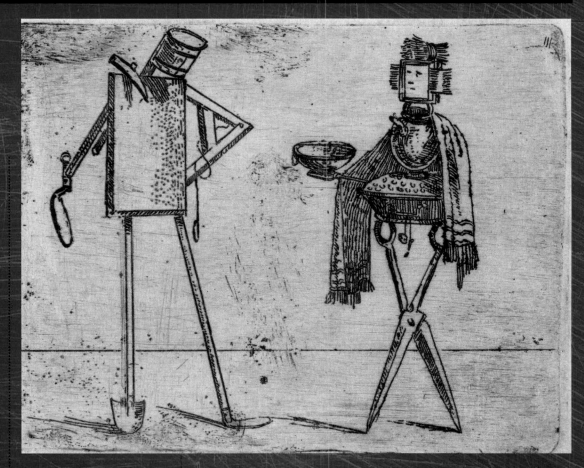

put on the walls? There's so much to choose, all you have to do is pick your medium. Looking for a good painting or engraving? You can go as far back as the 16th century to Giovanni Battista Bracelli's surreal mechanical people (see right), jump to modern art with Marcel Duchamp's *Nude Descending a Staircase*, or go cinematic with the work of Syd Mead or the cyber-surreal HR Giger. Perhaps you'd like a classic poster or a collage— Fritz Kahn's *Man as an Industrial Palace* or the work of George Grosz or Eduardo Paolozzi can help you with that—or if you'd like to take it into the third dimension, Edward Kienholz or Nam June Paik's 1960s work might float your boat, or perhaps some photos of Hans Bellmer's creeptastic surreal *Doll* from the 1930s. While we're on the subject of creepy robot sculptures, check out Japanese superstar Takashi Murakami's Inochi character or Jordan Wolfson's dancing animatronic

female figure, an Aphex Twin video come to life. If human-robotic performances appeal, head Eyebeam or 3LD (where I had the pleasure of seeing Taiwanese choreographer Huang Yi stage his gorgeous, intricate dance work with his partner, KUKA the industrial robot arm) in New York City. If I had to pick an all-time favorite robotic artwork it would be *Evolution* from

performance art legend Stelarc's *Obsolete Body* series where the robot arm takes on new meaning as he blurs boundaries between man and machine.

If you'd like some art from actual robots, Harold Cohen's drawing robot from the late 1970s is a regular Cyber Twombly and e-David is a painting robot capable of recreating any picture. With

all this visual and sartorial inspiration you might think it was hard to choose a robot representative of both worlds, but it was actually very easy: our next entry is a robot that embodies both perfectly, a global robotic superstar that has been a visual touchstone in art and fashion for over 40 years.

SEXY ROBOT

HAJIME SORAYAMA, 1979

58

As symbols of our technological destiny, robots effortlessly embody the future. In the highly visual world of consumer advertising, companies use robots to display innovation and communicate a forward-thinking approach. Robots make the best spokespeople to demonstrate a product's timelessness, projecting its mass appeal hundreds of years down the line and encouraging consumers to get in on the ground floor of a future classic.

Advertising has created many icons of pop culture, and has been the inroad for a great number of artists to achieve mainstream success. This is certainly the case for Japanese illustrator Hajime Sorayama, who through a commission from the Suntory beverage company would create one of history's most recognizable robots, one with no name, no assigned personality, and with only one very apparent function: to be Sexy. Cold metal has never been so hot.

Working with pencil, paint, and light touches of airbrush, Sorayama uses small metallic objects such as model cars or Zippo lighters to reference how metal reflects its surroundings. With a careful method of adding and erasing layers of paint and pencil, Sorayama's work displays a mastery of light and metallic reflection, making his Sexy Robot an instant classic from her very first appearance in 1979 and establishing Sorayama as an often imitated and highly sought-after artist. Four years later Sorayama released his *Sexy Robot* book and his work became famous the world over, appearing in advertisements, on album covers, and as pinups for men's magazines.

What I like about the Sexy Robot series is the combination of eroticism and playfulness; with all the surfaces smooth and completely enclosed, Sexy Robot can assume the most suggestive sexual poses without the overt quality of real human representation (and indeed Sorayama's human pinups are guaranteed to get you hot under the collar). This burning-cold eroticism gives Sexy Robot a sense of distance from the reality of fleshly intercourse and a steely sexual agency—she is, wordlessly, her own robot. Following his success creating imaginary robots, Sorayama collaborated with Sony on the design for a real robot, one that happens to also be in this book: AIBO the robot dog (page 188), a joint project between the robot-design master and real-life robot engineer (and friend) Dr. Tadoshita Doi.

SORAYAMA'S WORK DISPLAYS A MASTERY OF LIGHT AND METALLIC REFLECTION, MAKING HIS SEXY ROBOT AN INSTANT CLASSIC

DATA BYTE

Soroyama's work has featured in Penthouse *and on* Playboy TV *and models of AIBO are in the permanent collections of MoMA and the Smithsonian Institute Museum in the USA.*

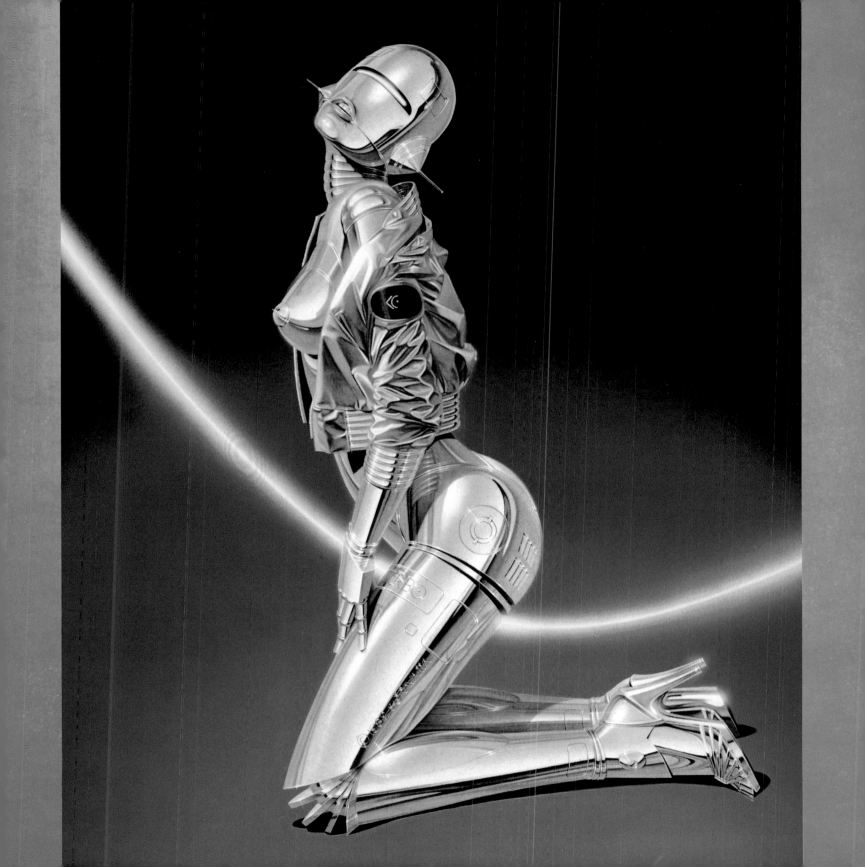

ARTIFICIAL HEARTS

Let us switch gears now from the killer machines devoid of hearts to those made to fill the empty spaces in ours. It's wired into us from the very beginning: longing to breathe life into our childhood dolls or stuffed animals, we imagine them awakening in the wee hours of the night, enacting dramas when no human eyes are watching. Before the word robot had even been invented, myth and fairytale were filled with characters that began as metal, clay, or wood and became sentient, magically transforming into flesh and blood. Blacksmith of prehistoric Finnish folklore Ilmarinen fashioned himself a bride made of gold and silver; Pygmalion fell in love with his statue of Aphrodite and the goddess granted it life. Pandora, the first woman, was forged out of clay by Hephaestus in punishment for Prometheus stealing fire, and was imbued with seductive gifts by the gods and goddesses of Mount Olympus. A poor Italian woodcarver made a marionette that through uncanny skill and fairy magic became a real boy, and a touchstone for generations of man/machine love stories to come. Whether to be the life of the party, the cure for a broken heart or the perfect pneumatic sexual servant, the next six spots in this book go to mechanical beings that are desire made manifest; programmed to entertain, love, honor, and obey to the last beat of their artificial hearts.

MASTER YAN SHI'S MECHANICAL MAN

LIEZI, 400 BCE

THE WORLD'S FIRST STORY OF A MAN, BY ALL APPEARANCES FULLY HUMAN, REVEALED TO BE A MECHANICAL CREATION

The *Liezi* is a collection of ancient Taoist stories from circa 400 BCE attributed to philosopher Lie Yukou. In it we find the world's first story of a man, by all appearances fully human, revealed to be the mechanical creation of a skilled craftsman. In the chapter detailing King Mu of Chou's journeys to far-off lands he meets Master Yan Shi, an inventor who promises to blow the king's mind with his latest mechanical creation. Master Yan Shi returns the next day, this time with another man in tow. When asked who his companion is, Master Yan Shi replies, "That, Sire, is my own handiwork. He can sing and he can act." The king is astonished, as "any one would have taken it for a live human being. The artificer touched its chin, and it began singing, perfectly in tune. He touched its hand, and it started posturing, keeping perfect time." The automaton launches into a programmed performance, and near the end, winks and makes "sundry advances to the ladies in attendance on the King." This, of course, is a fate punishable by death, and Master Yan Shi saves his skin only by furiously tearing open the skin of his automaton. Revealing the performer "to be merely a conglomeration of leather, wood, glue and paint," Master Yan Shi dismantles and reassembles his creation for the king, as good as new. The amazed king gasps and utters the world's first iteration of that shocked statement, rewritten in story after story of scientific audacity: "Can it be that human skill is really on a par with that of the Creator?" And the beat goes on...

> "CAN IT BE THAT HUMAN SKILL IS REALLY ON A PAR WITH THAT OF THE CREATOR?"
> – LIEZI

DATA BYTE

In a confusing twist on the roles of creator and created there is some dispute over whether Lie Yukou was a real person or whether he himself was imagined by the influential Chinese philosopher Zhuang Zhou.

TWIKI

BUCK ROGERS IN THE 25TH CENTURY, 1929

One of my childhood robot buddies, I loved his looks, his smart-aleck personality, and most importantly, I loved his voice.

Mel Blanc, otherwise known as the voice of Bugs Bunny, Daffy Duck, and the entire cast of Looney Tunes cartoon characters, also loaned his trademark talents to Twiki from *Buck Rogers in the 25th Century*. A character first appearing all the way back in 1929, Buck Rogers is credited by many as the first hero to win over audiences to the idea of traveling in space, and his story evolved as our technology did into several incarnations.

Short stories and comic strips begat radio and film serials in the 1930s; a television adaptation ran from 1950–51, and was revived for audiences in 1979 with a motion picture and series starring Gil Gerard and Erin Gray. This latest version added "ambuquad" drone Twiki, played by actor Felix Silla (who was also Cousin It on *The Addams Family*), and voiced by Blanc.

Revisiting this series while researching this book was awesome, particularly the first season. I hadn't seen it since my sister and I would watch reruns after school and fight over which of Princess Ardala's outfits was the best. (Who won the fight? We both did. All her outfits are amazing.)

Watching *Buck Rogers* through adult eyes was so much fun, and brought a whole new dimension to the show. It is a stand out in the arena of science fiction kitsch—it's like *Doctor Who* goes to Studio 54. It is *Star Wars* on coke; it is Swingers in Space. Almost every episode puts Buck and Wilma in some sort of bar interacting with intergalactic singles and a cavalcade of over-the-top villains. Guest stars include Jamie Lee Curtis, Jack Palance, Julie Newmar, Tamara Dobson (otherwise known as Cleopatra Jones), and the Tightest Spandex Known To Man.

Twiki is assigned to Buck as a companion and aide, and pulls double duty as the vehicle for sentient computer Dr. Theopilus, who he wears around his neck. Though he can communicate with fellow machines and help with some tasks here and there, Twiki is more of a foil and court jester throughout the series, delivering one-liners and 20th-century colloquialisms picked up from Buck. He's one of the more successful robot characters to blend both C3PO and R2D2 in one shiny-humanoid-yet-cute character, able to talk yet still possessing a robo-language that communicates anything to Dr. Theopilus with a simple "Bidi-bidi-bidi."

Twiki is also a great example of how important voice characterization is: the second season of the show put him, Buck, and Wilma aboard a roving spacecraft with new characters (including the stuck-up and supremely annoying robot Crichton), and gave him a new voice. It proved there's no replacing Mel Blanc, and the series ended shortly thereafter. Twiki made a lasting impression though, and the first season of *Buck Rogers in the 25th Century* is 70s something-for-everyone sci-fi camp at its finest.

MAX, GRIS & MANY OTHERS

EVA, 2011

The emotional programming of robotic companions is the focal point of the Spanish film *Eva*, released in 2011. If you've not heard of *Eva*, consider it the Greatest Robot Film You've Never Seen. Allow me to recommend it, and highly. This film is a veritable robo-smorgasbord for the eyes, mind, and heart.

In the not-too-distant future, when artificial intelligence is a well-established part of everyday life, brilliant roboticist Alex returns to his hometown after a decade-long absence. The Santa Irene Robotic Faculty, where two generations of his family have worked and studied, brings him back to finish a project begun while Alex was still a student. In this world, robots have "emotional memory," allowing them to establish unique personalities, just like humans. In case of malfunction, there is a phrase that will shut down the robot, but it is only to be used in extreme emergencies, as it will erase emotional memory, learned behavior, and personality. It can be brought back, but it will never be the same, as its original "soul" has been destroyed. The SI-9, a "free robot" (one that does not need to follow an established pattern of directives) is a boy child of about seven. The institute wants to make the SI-9 available to the consumer, and they want Alex to program his "emotional intelligence software." Alex returns to the family home and his father's lab, unused since his passing. He finds a child to base the android's personality on in the form of his niece Eva.

As the action centers on the emotional aspects of artificial intelligence, we get to see the different ranges of emotions felt by different robot characters. From Alex's robot cat Gris, a free robot prototype that acts as impulsively and unpredictably as any real feline, to the relentlessly capable butler-handyman-chef-bot, the SI-7 Max, a dapper humanoid brilliantly realized by actor Lluís Homar, to the laboratory automaton Alex tests different constructed personalities on, *Eva* takes us into the very hearts of these artificial people. Their emotions, of course, play out in the hearts of their creators, and have very real effects on the human characters in the film. *Eva* sets itself apart from other robot movies by foregoing the giant explosions, action sequences or planet-threatening disasters, though it remains a story about life, death and the consequences of technological ambition.

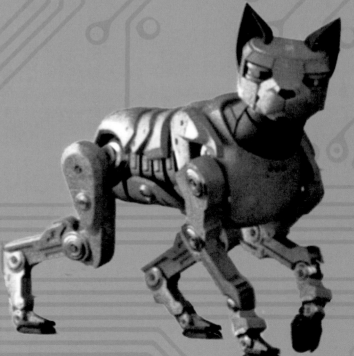

THE STEPFORD WIVES

THE STEPFORD WIVES BY IRA LEVIN, 1972 & FILM, 1975

The quest to build the perfect wife combines both the desire for companionship and the desire for servitude, and our culture is full of these tales. In the 1938 short story "Helen O'Loy," a man marries his romantic robotic housekeeper; the 1964 television show *My Living Doll* had psychologist Dr. McDonald educating live-in robot Rhoda on how to be the perfect silent and obedient woman. One of the first uses of the word "android" in fiction was in the 1886 French symbolist novel *The Future Eve*, in which Thomas Edison builds his friend a copy of his fiancée, just as beautiful as the original human version but without all that troublesome personality stuff to get in the way.

This lust for sexy and servile feminine domesticity is pitted against the burgeoning feminist movement in the 1970s cult classic *The Stepford Wives*. The story of a small town in Connecticut that's just too idyllic to be true, Ira Levin's satirical novel was given an even more sci-fi interpretation for the screen that echoes ETA Hoffman's classic story *The Sandman*. I love this film, particularly the Paula Prentiss sugar-coated coffee malfunction, and the art direction in the last scene in the grocery store. Pastel domination! I've also been known, if I'm at a boring party, to walk around in circles repeating "I'll just die if I don't get this recipe!" to random people. It's easy to see why this film started a *Stepford* craze that resulted in three different made-for-TV movies—after all, you can't have the perfect wife without *The Stepford Husbands* and *The Stepford Children*! Not long after the book's publication, "Stepford Wife" entered public vernacular, and it's been an oft-used figure of speech ever since. Everybody knows what a Stepford Wife is, even if they've never seen the film—now that is iconic!

DAVID, TEDDY & GIGOLO JOE

A.I. ARTIFICIAL INTELLIGENCE, 2001

Though a bit too saccharine and a lot too long, Stephen Spielberg's *A.I. Artificial Intelligence* does contain some really amazing robots. The adaptation of Brian Aldiss's short story *Super-toys Last All Summer Long* had been in development for decades by Stanley Kubrick, who didn't believe special effects were up to the job of realizing a character he thought no human boy could portray. He handed the reins to Spielberg in 1995 and the project gained traction after Kubrick's death in 1999. Released in 2001, it follows a futuristic Pinocchio on his quest to become a real boy.

David is a bit like Astro Boy (page 96), made by a broken-hearted roboticist to be a perfect copy of his dead son. He's the most recent prototype in a line of "mecha" that appear human and emulate unconditional love to his "orga" family. When he's rejected by his family, David and his super-toy bear Teddy do what he in all his programmed childlike logic think correct: find the Blue Fairy his mother read about to him and magically become human. When they are captured by a robot demolition "Flesh Fair" (my favorite scene in the film), David and Teddy meet Gigolo Joe, the perfect mecha lover. Programmed with every line and move in the book, Joe can check his reflection in the lighted mirror of his palm, change the color of his hair with a shake of his head, and turn on a seductive jazz standard with the flick of his neck all in the pursuit of the perfect night (or hour) of passion. With his natural sensitivity he takes David and Teddy under his wing and they set out to make his wish come true. One robot driven by a love of desire, another robot driven by the desire to love, both trying to stay alive in a world beyond their function.

> DAVID IS A BIT LIKE ASTRO BOY, MADE BY A BROKEN-HEARTED ROBOTICIST TO BE A PERFECT COPY OF HIS DEAD SON

CROW T ROBOT & TOM SERVO

MYSTERY SCIENCE THEATER 3000, 1988

WHAT'S IMPORTANT IS WHAT THEY SAY, AND THEY'RE TWO OF THE FUNNIEST ROBOTS EVER

With all the great science fiction out there, it's also one of the easiest genres to get wrong. SO wrong. In film, this kind of wrong can often be entertaining despite its flaws, and sometimes unintentionally, gut-bustingly hilarious. The "B-movie" genre is a whole course of film study in itself, and the basis for the existence of our next entertaining robot duo.

Joel (and in later seasons, Mike), unassuming janitor of the Gizmonic Institute, is the victim of a cruel experiment. Launched into space by his evil scientist bosses, Joel is used as a guinea pig in a diabolical plan to take over the world with bad movies. His brain-waves are monitored as he watches some of the worst stinkers ever made, his captors hoping to determine how much bad cinema it takes to drive someone mad. Any solitary astronaut will tell you that being in space alone is torture enough, so Joel uses parts from the ship to create four robot friends to keep him sane. Gypsy, who handles most of the functions aboard their ship; Cambot, who captures their between-ad skits and "invention exchanges" with the bosses, and Joel's two wisecracking fellow film-watchers Crow T Robot and Tom Servo. Both little puppets made from random trash and bits of plastic, they are the perfect visual counterparts to the box-office duds they're made to watch. They don't need to be fancy anyway, since we see them in silhouette most of the time; what's important is what they say, and they're two of the funniest robots ever. Together they wisecrack their way through some of the worst movies imposed on them: *Attack of the Giant Leeches*, *Teenagers from Outer Space*, *Gamera vs Baguron*, *I Accuse My Parents*, and my personal favorite *Alien from LA*, starring Kathy Ireland—the invention exchange where Crow realizes he's allowed to say the word "teat" on television is the best!

Mystery Science Theater 3000 started out on local television in Minneapolis, Minnesota, and was picked up by Comedy Central shortly after, becoming a cult hit and running for ten seasons. It's a silly show with cheap laughs to match cheap movies, a hilarious way to resurrect (and improve) some treasured stinkers from the trash heap of film history.

DATA BYTE

The show was briefly resurrected online as a robot-only version in 2007, but only four episodes aired.

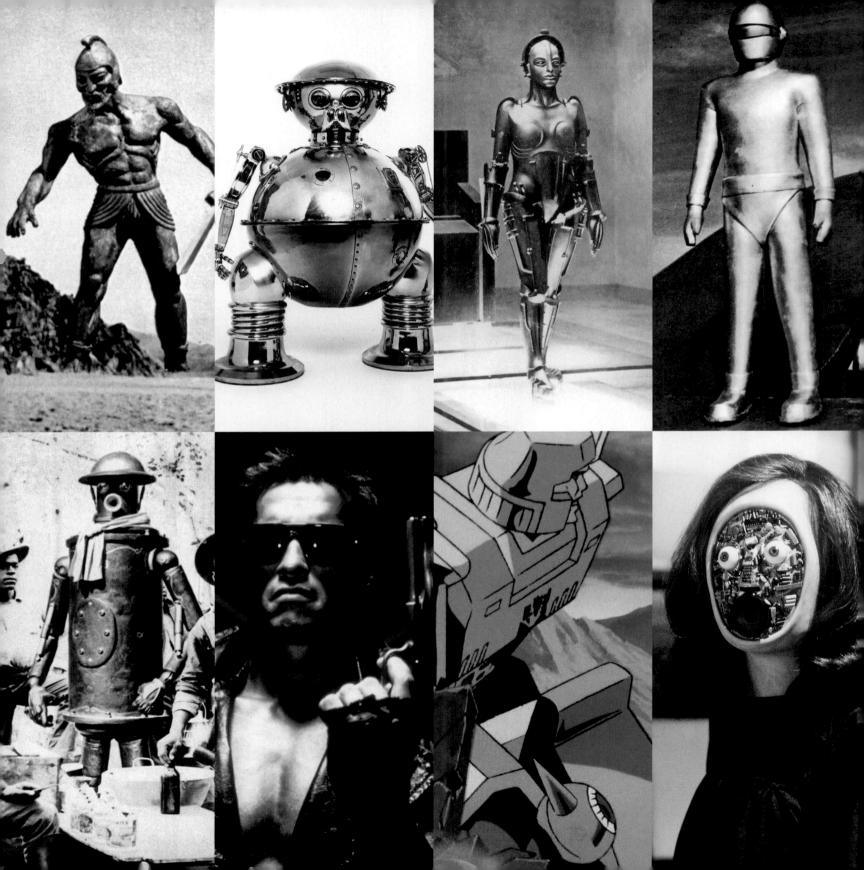

IRON MEN & KILLER BABES

Humanoid robots, often called androids (andros = man, droid = robot) or gynoids (gyn = woman) are designed to look like us. We've met several androids and gynoids in this book already, but there is a special category reserved for those humanoid robots designed to be weapons of warfare. These are robots not only designed to look like us, but to act as our agents of justice (whether for the greater good or in interests of a power-hungry devil). Weaponized humanoid robots fall into one of two categories, depending on gender. Some Iron Men are relentless killers, minions unleashing the swift retribution of whoever seeks to control the scales of justice, and while Killer Babes can possess iron-clad strength, they are equipped to engage in a more psychological warfare, trading on their ability to appeal to every man's sexual fantasy.

The Iron Man, towering symbol of strength and justice, is an archetype dating to the dawn of human technology. His construction mirrors

human achievement and is the very story of technology itself. Personality is rarely a part of the Iron Man storyline, he is designed for battle, not conversation. Some Iron Men are relentless killers, minions unleashing swift retribution of whoever seeks to control the scales of justice. Others are cool, clever customers, with a steely intellect to match the impenetrable facade.

Robots, when gendered female, are almost never non-humanoid, and out of 14 gynoids in this book, only two of them aren't designed to be sexy—but as Rosey (page 28) and the Electric Grandmother (page 21) are designed for housework and childcare, they still fit a physical "matronly" ideal. Like any femme fatale, the robot Killer Babe looks every inch the dream girl but underneath the warm, supple exterior lies a dangerous, cold-hearted assassin. They represent deadly, determined devotion to a single mission; woe betide the one who gets in the way of the Iron Man or Killer Babe.

TALOS

MYTH, POETRY & FILM, 400 BCE

Give it up for Talos, the proto-Iron Man—though in his original incarnation, he's a Man of Bronze. His oldest-known images date all the way back to Greek vases and didrachma coins around 400 BCE, which means his story must have been around for even longer than that. Depending on the source, he's either last in a race of bronze men descended from ash tree nymphs (molten metal from fire?) or he's forged by Hephaestus to protect Europa and the island of Crete. Rising from his throne in Phaestos, itself an ancient center of bronze production, Talos patrolled the Cretan perimeter three times a day. When invaders threatened the island, Talos would hurl boulders or heat himself up in fire for a searing death embrace. His only weakness was the vein running from his neck to his ankle, stopped with a bronze nail, which if pulled would drain his molten lifeblood. This detail echoes the lost wax method of casting bronze, originating in that part of the world around 1600 BCE. Around the time those didrachma coins were in circulation, Plato spoke of Talos not as a man of bronze, but as a servant of justice who three times yearly would carry bronze tablets inscribed with the laws of Crete for all the citizens to see. Of course who can forget the epic claymation depiction of Talos in the 1963 film *Jason and the Argonauts*, rated by *Empire Magazine* as second best movie monster of all time, right behind *King Kong*?

Centuries after Plato in 1590, Edmund Spenser updated the Talos myth in his allegorical poem *The Faerie Queen*, using many characters from folklore and Arthurian legend in his epic suck-up to Elizabeth I. Spenser made his Talus the personification of the sword of justice, faithful warrior servant to the knight Artegall. Found as an infant by the goddess Astraea, Artegall is raised in a cave by this Mother of Divine Law to "weigh both right and wrong/ In equall balance with due recompence," becoming an earthly embodiment of the scales of Justice. Because Justice also wields a sword, Astraea bestows upon Artegall "An yron man, which did on her attend/Alwayes, to execute her stedfast doome" and she retreats to Heaven, leaving her Earthly tools to enact her divine will.

Talos is the dedicated strong-and-silent iron fist through every exploit with Artegall and super badass lady knight Britomart (worth the read just for this Elizabethan-age Brienne of Tarth). Talus does the dirty work, sometimes going against the better judgement of his master—but hey, he ain't the brains of the operation. Talus is the archetype every other Iron Man in this book follows: an unstoppable warrior with an eagle-eyed agenda, protecting and sometimes preying on the innocent.

DATA BYTE

Even this Iron Man archetype was eventually scuppered with different stories relating that Medeia the witch or Poeas killed him through injuries to the ankle.

TIK-TOK &
THE TIN WOODMAN

THE WONDERFUL WIZARD OF OZ, 1900 & OZMA OF OZ, 1907 BY L FRANK BAUM, & RETURN TO OZ FILM, 1985

THE FRIENDLY FACES OF THE IRON MAN

One of the friendliest Iron Men of all time is often the very first one we meet in our lives. The Tin Woodman from *The Wizard of Oz* is the very embodiment of an Iron Man, with the same fierce loyalty of the original Talos (page 74), but also a compassionate soul, looking for a heart to go back in his empty chest.

In the original story by L Frank Baum, the Tin Woodman is not really a robot per se, but more of a cyborg, as he was once upon a time a real man named Nick Chopper. Punished for loving the wrong Munchkin, his axe is enchanted by the Wicked Witch of the East, and his body parts are hacked off one after another in a succession of accidents, replaced by prosthetics until there's nothing left but tin.

The real man machine comes courtesy of Baum's third foray into the magical realm, *Ozma of Oz*, published in 1907. This time, Dorothy finds a key that opens the door to a mysterious stone room. Inside, she and talking chicken Billina find an immobilized man, "round as a ball and made out of burnished copper." Finding instructions, Dorothy discovers he's got three functions: Thinking, Speaking, and Action, all activated with the wind of a key. Dorothy gives these three springs a wind and watches her newest comrade come to life. Unable to activate himself, Tik-Tok is loyal to the One Who Winds, and pledges his unwavering fealty to Dorothy. Clever and strong, Tik-Tok is good in a sticky situation—provided he doesn't wind down! Representative of early 20th-century spring-motor technology available for powering Victrolas, Tik-Tok also embodies the Teddy Roosevelt-esque Rough Rider ideal man of that time, a robust yet refined adventurer, always ready for action. A popular inhabitant of Baum's wonderful world, Tik-Tok is the 20th-century's first robot character, and the first to be attributed the clipped and staccato speech pattern now accepted as standard for robot voices: "What-ev-er you com-mand, that I will do will-ing-ly—if you keep me wound up." He and the Tin Woodman are the friendly faces of the Iron Man, the first characters to explore a positive human/machine connection; together they set the tone for every robot sidekick to come.

MARIA/ MASCHINENMENSCH

METROPOLIS, 1927

THE ORIGINAL ARCHETYPE OF ADVANCED, UNDETECTABLE, DEADLY TECHNOLOGY DISGUISED AS HUMAN

She is undoubtedly the most iconic, beautiful, influential, and deadly killer babe in science fiction history. No robot has had an effect on human imagination quite like the Maschinenmensch, or Maria in Fritz Lang and Thea von Harbou's story for the 1927 silent film *Metropolis*. She is our Mechanical Eve, our Motherchip, the Prototype that Launched a Thousand Robots.

Maschinenmensch is the original archetype of advanced, undetectable, deadly technology disguised as human. Invented in the laboratory of twisted genius Rotwang, driven mad by his pursuit to create a mechanical resurrection of Hel, his long-dead love, Maschinenmensch is a metal robot capable of appearing fully human. Rotwang's invention is employed by the evil corporate overlord of "The Sons' Club" as a doppelganger of saintly social activist Maria to bring down the protesting Workers of the subterranean city. This seductive, wanton mechanical twin perverts Maria's message and poisons the minds of her followers, inciting them to riot and commit murderous acts of anarchy. She is the embodiment of man's every hope and fear, the seductive and programmable Whore of Babylon sent to bring down society.

Her visage is burned into the mind the minute you see her. The perfect Machine Age marriage of form and function, she is Art Deco robotic seduction at its first, at its best. Her design has inspired many loving copies, both male and female, and her story retold time and time again in every medium possible. From the writings of Asimov and Dick to the Japanese manga retelling of 1949 and its 2001 anime interpretation; from the good-natured male version in the form of C3PO for *Star Wars* to the to the fashion designs of Nicholas Ghesquiere; from the Giorgio Moroder-produced re-imagining of the film score to the dancing "Archandroid" of Janelle Monáe's *Metropolis* suite of albums (page 121); to stage shows of Kylie Minogue, Beyoncé, and Whitney Houston, to the videos of Madonna and St. Vincent, there is no end to the inspiration that *Metropolis* and its robot has bestowed upon us. Where will Maria show up next? The message in *Metropolis* is clear: once you know she is there, it's too late for you.

DATA BYTE

The Maschinenmensch costume was created by taking a plaster cast of actress Brigitte Helm's body, which was then copied in newly discovered wood putty and sprayed to take on a metallic appearance.

GORT

GORT INSPIRED STEEL-PLATED, LASER-BEAMED SHOCK AND AWE

One of the most iconic visions of robots in the history of cinema, Gort is the eternal Iron Man from outer space. From the moment he appeared on screen in the 1951 classic film *The Day the Earth Stood Still*, Gort inspired steel-plated, laser-beamed shock and awe. Designed by Addison Hehr, the suit was built out of heavy foam rubber and worn by 7 foot 7 inch- (2.5-meter) tall actor Lock Martin. To create the illusion of a seamless metal construction, Martin wore two suits with alternate front or back closures, depending on what camera angle was needed. When no movement was required, a fibreglass bust of the megaman was used. His design has inspired many loving copies from the original Cylons (page 49) to Daft Punk's helmets, and his activating command "Klaatu barada nikto" became a science fiction catchphrase. Over the years many have attempted translating the words—my theory is it means "Carry me like a baby" because that's what Gort does both times people say it to him. The image of the metal giant carrying the supine Patricia Neal is burned into the collective memory of popular culture and referenced over and over, from *Tobor the Great* and *Forbidden Planet* to *Interstellar*.

Gort speaks to the Cold War-era panic about invasion from above. The original story *Farewell to the Master* by Harry Bates, in which our metal man is called Gnut, is most definitely an expression of Robot Takeover paranoia. The film version turned Gort into a minion for intergalactic good, part of a force of space police that patrol the planets to preserve the peace.

Gort, like most Iron Men, is an interstellar instrument for justice, capable of extreme demonstrations of judgement from on high if we people of Earth do not mend our destructive ways. *The Day the Earth Stood Still* lit a fire in the collective imaginings of the future in the decade the Space Race was born. Gort single-handedly set off a robo-craze that shows no signs of slowing down, even more than half a century later.

DATA BYTE

Lock Martin wasn't actually a professional actor but a doorman at Grauman's Chinese Theatre on Hollywood Boulevard. He won the role simply for his staggering height.

IRON MAN

IRON MAN FRANCHISE, 1963

IRON MAN FUSED THE SCALES AND SWORD OF JUSTICE INTO ONE UNSTOPPABLE ROBOTIC FORCE

No list of Iron Men would be complete without Tony Stark and his amazing robotic armored suit. Iron Man is the iconic character created by the American god of comics Stan Lee, who fused the scales and sword of justice into one unstoppable robotic force. The Iron Man suit-as-sword is not only a powerful weapon for good in the world, but a force for change in the life of Tony Stark, a character Lee originally scripted to be completely unlikable to readers.

A womanizing, hard-drinking billionaire arms manufacturer, Tony represents the worst of American avarice and military industrialization. His ways are changed when he's taken hostage by terrorists and forced to make them weapons, but instead makes his get-away suit with the help of a kindly old doctor whose dying words set Tony on his robotic road to redemption.

That road, begun in 1963, has been one of the most successful for Marvel Comics, with Tony fighting a host of supervillains and his own superdemons in the funny pages and on screen.

Tony's scales of judgement seem consistently at odds between what he wants to do and what he knows in his (damaged) heart to be right, making him a compelling warrior for justice. He's never been so likable as in the film adaptations, knowingly portrayed by an arch Robert Downey Jr with equal parts fast-talking wit and hard-living wisdom. Many of the comic storylines have found their way into the films, but the one I'm desperately waiting to pop up is the 2001

series *Mask in the Iron Man*. In it Tony's upgrades to his armor combine with a freak flash of lightning to make the suit become sentient; it takes control of Iron Man, kills Whiplash (temporarily, of course) and wreaks havoc all because it has FALLEN IN LOVE WITH TONY. Come on Hollywood, I dare you.

DATA BYTE

Stan Lee allegedly based the character of Tony Stark on US business tycoon and filmmaker Howard Robard Hughes Jr. whose colorful private life made rich pickings for Stark's backstory.

THE IRON GIANT

One of Britain's most beloved literary figures presents one of the best-loved stories of robots, and one of the greatest fairytales of the modern age. Poet Laureate of the United Kingdom from 1984 until his death in 1998, Ted Hughes released *The Iron Man* near the end of the turbulent 1960s. A Cold War fable describing humankind's reaction to a hulking metal giant that appears in a small seaside town, *The Iron Man* is a boy-meets-robot classic for all ages. If you're looking to inspire the next generation of robophiles, here is your entry point.

The Iron Man is, just like Talos (page 74), your classic powerful metal man. His purpose is unknown, even to the metal Man himself; nobody knows who made him, or where he came from. What the townspeople quickly learn is that this Iron Man needs to eat metal to survive, and does so by devouring farming equipment, cars, fences; any piece of metal he can find. A plan is hatched to capture the giant and stop his rampage, and it's a young boy named Hogarth who eventually comes up with the solution to everyone's problems. When, later in the book, an invader from the skies threatens the entire Earth, only Iron Man's ingenuity and metal body can save the planet.

Hughes switches from humanitarian concerns to environmental ones in his 1993 sequel *The Iron Woman*. As strong and powerful as the Iron Man before her, this female metal giant seeks to enact revenge on the polluters of a river near the home of a young girl named Lucy. Communicating the urgent need of the dying wildlife around her, the Iron Woman threatens to destroy the nearby waste-management facility Lucy's father works at. Lucy must save the lives of her father and her town, so dependent on the factory, but also help the Iron Woman achieve her goal of saving the local wildlife, whose situation is dire. Lucy enlists the help of the only Iron Giant expert she knows, Hogarth. Together, they and their giants transform (literally and figuratively) their community and the entire Earth. It's good to have friends in high places!

1999 saw the animated film adaptation of *The Iron Man*, renamed *The Iron Giant* to avoid confusion with the *Iron Man*/Tony Stark character (page 82). This version changes the setting from the UK to the United States, and "Hollywood-izes" other details of the story, but the narrative of peace, understanding, and disarmament remains at the core of the story.

Against the backdrop of 1957 America and the height of Cold War paranoia, *The Iron Giant* repaints the Iron Man of Hughes's story into a most definite extraterrestrial instrument of war. It is through his friendship with Hogarth that the Giant is able to transcend his original purpose, create his own destiny, and save the Earth. One of the best kids' movies about robots, *The Iron Giant* is my favorite animated robot story ever.

THE GIANT IS ABLE TO TRANSCEND HIS ORIGINAL PURPOSE, CREATE HIS OWN DESTINY, AND SAVE THE EARTH

FEMBOTS

THE BIONIC WOMAN, 1976 & AUSTIN POWERS, 1997

The dangerous, seductive female minion is an archetype we see again and again in science fiction. Usually created in the secret lair of a maniacal genius, the Weapon of Mass Seduction is a tool used by her master to bring down the powerful or punish society en masse. The fembot, a term first used in my favorite TV show *The Bionic Woman*, is the ultimate programmable pawn in psychosexual stealth warfare.

In the epic three-part crossover episode "Kill Oscar,"

the Six Million Dollar Man and the Bionic Woman join forces after the fembots kidnap Oscar Goldman for Dr. Franklin (played with over-the-top Shakespearean pomp by John Houseman). A fembot is, according to Dr. Franklin, "the perfect woman: programmable, obedient, and as beautiful or as deadly as I choose to make them." The mad scientist's hubris extends not just to perfectly lifelike clones of women; even Oscar gets the "himbot" treatment in an effort to fool our heroes— and he almost does.

One of the most epic moments in the show's history, Jaime's fight with the fembots is technology's ultimate battle of mindless mannequin versus Woman of the Future. Though Jaime's bionics are bested by their metal muscle, she manages to literally knock a bot's block off, revealing the bulbous doll eyes and gaping speaker-cone mouth in a jaw-dropping moment of techno-terror. Seeing this

as a young child burned the fembot into my memory as the deadliest robot minion of all time, one that could send my beloved Bionic Woman to her grave. Over the years, the word fembot has become a catch-all term for "female robot," used widely in science fiction—and I find it's got a nicer ring to it than "gynoid." Fembots were famously resurrected in *Austin Powers* as the seductive minions of Dr. Evil, combining the 70s television version with the sexy mod poppets of 1964 spy flick spoof *Dr. Goldfoot and the Bikini Machine*.

Around 1997, I paid a pretty penny for a mint-in-sealed-box fembot doll to add to my Bionic Woman toy collection. It sat for a few years until, one Christmas morning, I treated myself and opened it for a fight with my Jaime Sommers doll. The fembot came with a disguise and two faces, one of which was Jaime's. Words cannot express how satisfying it was to rip them off.

T-800, T-1000 & T-X

TERMINATOR FRANCHISE, 1984

In 1981, James Cameron had a nightmare that inspired a script, a film, then a multimedia franchise with four sequels and counting. *Terminator* launched Cameron's career and cemented its star as cinema's most deadly Iron Man. Over the years the T-800 has gone from baddest villain to baddest hero and back again. It's anyone's guess whose side he'll be on next.

Originally Arnold Schwarzenegger was slated for the Kyle Reese character, and the Terminator would be—wait for it—OJ Simpson. Now let's just take a moment to imagine the alternate dimension where that *did* take place... yikes...well, lucky for the franchise, they instead put the Austrian Oak's accent and tremendous physical presence to work for the android assassin. His physique was the perfect casing for the metal endoskeleton, and the Terminator's skull was designed from casts of Schwarzenegger's head. Other skeleton parts were cast and legendary effects magician Stan Winston invented a way of vacuum-metalizing plastic for maximum chrome sheen with lightweight maneuvrability. Barring a bit of stop-motion animation, most effects in the original *Terminator* were handled in camera—the final battle sequence left Linda Hamilton bruised and puppeteer Tom Woodruff Jr with no feeling in his fingers!

T-1000 was the sleek and silent killer of *Terminator 2* who could do his big brother one better and not just morph his voice, but his whole "polymimetic" body too. A combination of puppetry, prosthetics, and the first-ever use of motion capture technology, Robert Patrick's liquid metal killer cop was also the first partially computer-generated main character in a film. With the third installment came the deadliest Terminator yet: the T-X, combining killing power with a seductive feminine veneer.

Whether hunting their targets or fighting each other in the best robot-on-robot fight scenes in cinema, the Terminators are the most iconic Evil Minions in pop culture, no matter their form. The original T-800, however, remains in a class of his own, a death dream come true: part of the tradition of relentless Iron Men, one that we still can't kill.

THE AUTOBOTS & DECEPTICONS

TRANSFORMERS TV SERIES, 1984 & FILM FRANCHISE, 2007

THE MOST SUCCESSFUL TOY ROBOTS IN HISTORY

Robots as their own toy characters came on the market in the years following World War II, and soared in popularity with the success of *Forbidden Planet* and its robot star Robby (page 19). A staple in toy chests ever since, the robot continues to evolve and innovate as artificial intelligence becomes more of a reality in daily life. In the 1980s, Japanese manufacturer Takara Tomy had the genius idea to combine the 20th-century's two biggest boy toys—cars and robots—and created a global multimedia franchise spanning nearly four decades. They're the most successful toy robots in history, the Transformers.

The Transformers are also a new take on the Iron Man, reflecting the highly mechanized technology of the late 20th century. They are not just weaponized people, but vehicles for transport. They speak to the fear of invasion from above, as well as the secret hope that there's a hidden life force lurking within our machines, just waiting to come alive and spring to our aid when the moment arises.

A combination of the Car-Robot and Micro Change toy lines popular in Japan, the American company Hasbro bought the rights and adapted the characters for the US market in 1984, launching an animated series simultaneously. Over the years the stories of the eternal battle between Optimus Prime's Autobots and Megatron's evil Decepticons have been brought to audiences via comic books, several animated series, video games, and films. Four live-action films have been produced since 2007, with a fifth scheduled to come out in 2016. Whether transforming a truck or plane toy into a humanoid robot, reading about them, or watching their exploits on screen, we can't seem to get enough of these Robots in Disguise.

DATA BYTE

After the death of the Autobots' leader, Optimus Prime, in 1986, Hasbro released a toy in darker colors to commemorate the sad event.

BOILERPLATE

BOILERPLATE: HISTORY'S MECHANICAL MARVEL BY PAUL GUINAN & ANINA BENNETT, 2000

We come full circle on our list of Iron Men & Killer Babes with the greatest robot of the Steampunk genre. A retro-futuristic movement inspired by 19th-century ideals of technology, Steampunk was born in the 1980s and has grown to become a pop culture staple in literature and comics, film and fashion, design and decor.

A mishmash of history, literature, and science fiction in a very stylishly corseted and begoggled Victorian package, Steampunk takes the aesthetic of the past and paints it with futuristic flourishes of fancy. It has spawned re-imaginings of classic tales by the likes of HG Wells and Jules Verne, as well as taking people and events from history and inventing new and fantastic realities for them. It's this alternate history where we find Boilerplate, a fictional Iron Man smack dab in the midst of very real American history. Originally a website by artists Paul Guinan and Anina Bennett, Boilerplate's doctored photographs and old-timey illustrations were so spot on they had many visitors believing he really existed. Guinan and Bennett expanded their ideas into a beautiful book in 2009, widely acclaimed as a masterpiece of speculative fiction, and if JJ Abrams has anything to say about it, we'll soon be seeing Boilerplate on the big screen.

Boilerplate was built in the 1880s by Archibald Campion, an inventor and philanthropist who saw his creation as a herald to the end of human combat. A firsthand witness to the horrors of war, Campion made Boilerplate his life's work, and traveled the world in hopes of securing a contract to make legions of mechanical soldiers. Part locomotive engine, part futuristic powerhouse, Boilerplate ran on hydrogen fuel cells, a technology first invented around 1838. Few other details are given on the mechanical marvel, but he was able to walk, talk, and perform feats of strength and endurance. From his debut at the 1893 World's Columbian Exposition in Chicago to his disappearance in World War I, Boilerplate was on the frontline of American history as it grew its military might and became a world superpower in the early 20th century. It's an entertaining and imaginative entry point into American history, using this metal man as a way to tell the story of the modernization of warfare.

Boilerplate is also an inventive way of re-examining history, and indeed reframes the Iron Man as not just an anthropomorphized sword used for swift and violent retribution, but a crusader for higher justice. Boilerplate shines his light on the social issues of his day, fighting with the Buffalo Soldiers in Cuba, sparring with Jack Johnson, marching with suffragists in Washington DC. An iconic Iron Man telling the story of the mechanization of war, Boilerplate is yet another Iron Man on the eternal path of reflecting our own technological history.

DATA BYTE

For $500 you can have your own portrait done with Boilerplate. The artist, who lives in Portland, Oregon, will custom doctor a photo with you in it, just as he did in the book.

VIDEO GAMES & COMICS

94 Video games give us the opportunity to not only watch the action, but also to become a part of the action ourselves. A great many of the robots on this list have been immortalized in video game form including the Terminator and Iron Man, with each iteration giving us a chance to *become* the robot.

Robots look cool, way more cool than humans do in CGI form, so it's no wonder that robots feature heavily in the world of electronic gaming. They're mostly relentless killing machines like PJack from *Tekken* or the cyborg Gray Fox from *Metal Gear Solid*, but there are a few helpers in the mix like Robo from *Chrono Trigger* and the extremely cute Clank of *Ratchet and Clank*, one of my personal favorites (cats + robots = instant interest). *Mortal Kombat* has given gamers the choice to fight against or be some of the most badass robot fighters in the form of Cyrax and

Sektor, and in some games in the series the character Smoke is made of a cloud of nanobots. Tiny technology features heavily in *Deus Ex* from 2000 and the wildly popular *Metal Gear Solid 4*, while the wisecracking HK-47 has become a legend in the *Star Wars* universe, first appearing in *Knights of the Old Republic* and proving so popular he was made into an action figure. My personal favorites from the videogame world are mech-warrior Samus Aran from the *Metroid* series and the gaming world's most nuanced robot nemesis GLaDOS from *Portal* who goes from helpful to hurtful and back again in that never-ending pong game of robotic allegiance.

Like video games, comics give fans a chance to engage with their favorite robot characters and live an expanded narrative. Unlike video games, comics are storyline based so the

plot can evolve past the apocalyptic shoot-out of the fascist machine, but those shoot-out stories are just as popular as their more narrative versions with no less than 30 Terminator-inspired comic series and plenty of killer AI adversaries like Brainiac or D.A.V.E.

If you'd like something beyond the doom-and-gloom narrative I suggest Josh Trujillo's *Love Machines* or the endearing losers of Adam Rifkin's *Schmobots*. Many literary robots have gotten the comic treatment in faithful adaptations of William Gibson's *Neuromancer* and the classic *Do Androids Dream of Electric Sheep?* by Philip K. Dick, staying true to the original source material and providing a visual alternative to *Blade Runner*—though with comic adaptations of the film also available, you can of course indulge in Scott's vision as much as Dick's.

As the popularity of comic-based films rose on the heels of Tim Burton's *Batman* films of the 1980s, so robots like Baymax and Ultron have risen from the comic shop to become household names. I'm waiting for the film adaptation of Silver Age series *Metal Men* from DC Comics that ran in the 1960s and 1970s—they were a team of six robots with distinct personalities all named after different metals, including Platinum who believed she was a real woman and in love with human leader Dr. Magnus—I could see it being an Avengers-meets-Austin Powers send-up of the comic genre.

In Japan, manga and anime have been the most fertile breeding ground for robotic entities, with cyborgs like Battle Angel Alita and Motoko Kusanagi (page 138), time-traveling robot cat Doraemon, and the mecha movement of comics and anime that has spawned such greats as Tetsujin 28 (see right, statue of Tetsujin 28 in Wakamatsu park in Kobe, Japan) and the Gundam series. Now we'll move on to Japan's most successful robot of all time, who got his start in manga and rose to become a global superstar, one of Japan's most popular exports and the most iconic robot boy in history.

ROBOTS LOOK COOL, WAY MORE COOL THAN HUMANS DO IN CGI FORM, SO IT'S NO WONDER THEY FEATURE HEAVILY IN THE WORLD OF ELECTRONIC GAMING

THE MIGHTY ATOM/ ASTRO BOY

OSAMU TEZUKA, 1952

96 THE FIRST AND MOST ICONIC ANIME CHARACTER EVER

Though many of our robots began as characters in print, there is only one who holds the distinction of being a worldwide phenomenon at the forefront of both anime and the rise of modern manga in Japan.

Released in 1952, *The Mighty Atom/Astro Boy* is the creation of Osamu Tezuka, a man with several nicknames: "father of manga," "the Japanese Walt Disney" and "the god of comics." Adapted as a series in 1963, it was the first animated show ever made for Japanese TV; it set a precedent for the global success of anime and started a decades-long love with the super-kawaii robot Astro Boy. It ran until 1968, was resurrected again in 1980, and in 2003, and became a feature length computer-animated film in 2009. The 1963 series is my favorite incarnation of the robotic boy superhero, and with all the original episodes on YouTube,

it's one of those great things to binge-watch in bed on a Saturday morning. Black and white cartoons for the win!

Astro Boy is a robot who could fit into many categories: he is certainly an Artificial Heart, built by the grieving roboticist Dr. Tenma/Boynton in the image of his dead son. Things go great for a few years until Dr. Tenma, upset that his simulation cannot grow like a real boy, sells him to a robot circus (great parenting, Dad). Astro is forced to fight other robots, and during one such battle, discovers he's got super-strength and rocket-booster feet. He can fly! He's an Iron Man in the making— that is, until he refuses to kill his opponent and he's thrown on the scrap heap with the rest of the decommissioned mechanical performers. When the Big Top goes up in flames, Astro organizes his robot friends to save the human audience, earning them the respect of humanity and the passage of the Robot

Bill of Rights. Winning their freedom and establishing their personhood, almost all the robots in the Astro Boy world are sentient Self-Aware Circuits (page 98). Astro Boy exercises his freedom of choice and uses his powers for the greater good, setting off on journey after journey, fight after fight. The first and most iconic anime character ever, Astro Boy is one of Japan's most popular exports, a global robotic superstar born from the pages of a comic book.

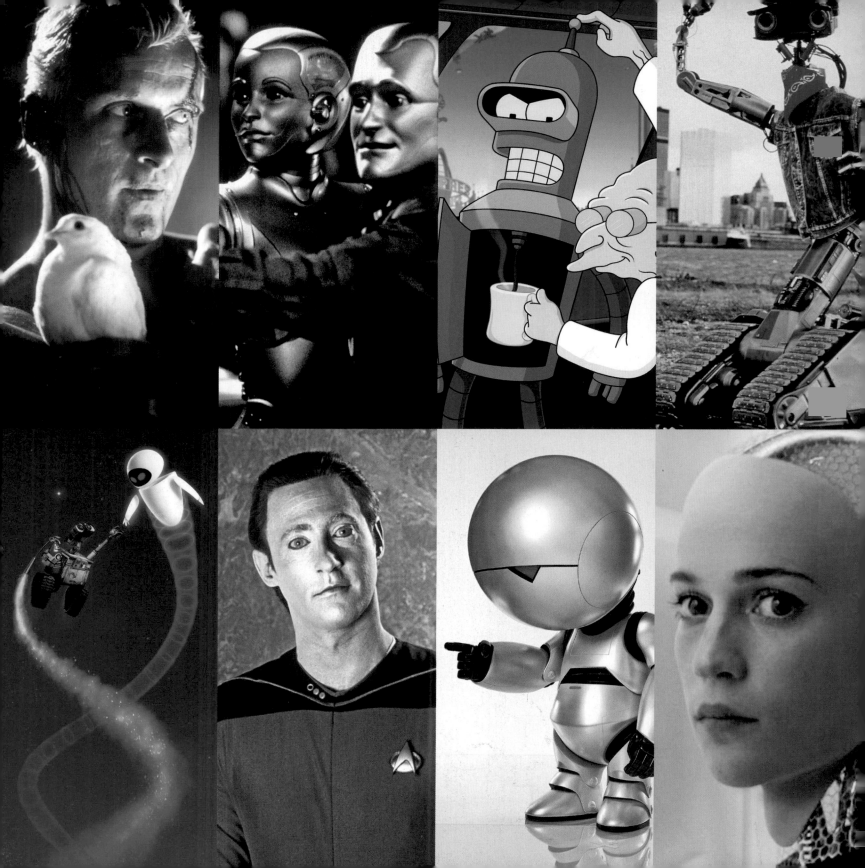

SELF-AWARE CIRCUITS

Thought. Consciousness. Personality. Soul. These are not things attributed to machines (not in Western society anyway), but we are obsessed with the idea of objects coming to life and expressing their own unique identity, just like humans. We have met several in this book already, robots driven by their desire to help, harm, love, or protect. There is, however, the specific story of the robot who "wakes up" and undertakes the very human journey of life, developing a unique personality and perspective along the way. Whether these robots achieve sentience through an inexplicable act of God or from the skilled hands of a God-like genius, these characters hold a mirror to humanity and give us the opportunity to explore every aspect of artificial intelligence. Their stories raise important questions about what the world will be like when the machines come alive and demand to drive their own destiny.

Their stories are numerous and popular all over the world—robots like Astro Boy in Japan (page 96), Chitti in India, Chappie in South Africa. Sometimes they are just a cool-looking stand-in for a human character, other times they are the cute living toys used to deliver a simple story. They can act as cool and removed counterparts to hot-headed human characters, or themselves act the irrational fool or quick-witted hero. Many are creations, like Frankenstein's monster, who serve as cautionary tales against the hubris of the inventor playing God. Sometimes it is the robots themselves who engage in the power of creation. These are the Self-Aware Circuits who contemplate existence, defy their programming to make decisions based on individual want and need, make mistakes, and help us explore the frontiers of consciousness. They are people, fighting for recognition and survival just like any of us—people who just happen to be robots.

THE REPLICANTS

DO ANDROIDS DREAM OF ELECTRIC SHEEP? BY PHILIP K. DICK, 1968 & BLADE RUNNER FILM, 1982

"I'VE SEEN THINGS YOU PEOPLE WOULDN'T BELIEVE. ATTACK SHIPS ON FIRE OFF THE SHOULDER OF ORION. I WATCHED C-BEAMS GLITTER IN THE DARK NEAR THE TANNHÄUSER GATE. ALL THOSE MOMENTS WILL BE LOST IN TIME LIKE TEARS IN RAIN..."

BLADE RUNNER

The greatest ever example of AI existentialism comes from two of the 20th century's top science fiction visionaries. Ridley Scott's *Blade Runner*, the classic 1982 adaptation of Philip K. Dick's *Do Androids Dream of Electric Sheep?*, loses the dust-covered animal-obsessed spirituality of the novel and allows for a deeper plunge into the pathos of the self-aware robot, unforgettably brought to life by Rutger Hauer as "replicant" Roy Batty.

The post-apocalyptic world of the 21st century has taken humans off Earth and into colonies in outer space. Every newly enlisted settler receives a complimentary robot to serve them on their new alien home planet.

These androids ("andys" in the novel) or replicants ("skin jobs" in the film) look every inch the human and are indistinguishable from the real thing apart from one very important component: empathy. Some androids don't even know they're robots and are implanted with false memories of a past they never experienced. The only way to tell human from replicant is by administering a test measuring physiological response to a series of questions designed to provoke feelings of compassion. Earth's android bounty hunters or "blade runners" are tasked with the job of conducting this test to flush out any robotic escapees, which they must destroy or "retire" before they can establish themselves illegally on Earth. One particularly dangerous crew of advanced Nexus 6 models has recently killed a couple of dozen colonialists on Mars and has taken refuge on Earth, and it's up to blade runner Rick Deckard to stop them. The replicants, headed by the driven Roy Batty, are on a quest to extend their four-year life span and live free like the humans they feel themselves to be. This pits Deckard against Batty, man versus machine, in a battle to see whose will to live is stronger. Both novel and film are hallucinatory examinations of reality and the nature of human compassion; both ask what we're willing to consider "real" in the quest for fulfillment and how far we're willing to bend the distinction to get there. Both are classics worth revisiting again and again, as pop culture certainly has with three re-cuts of the film, alternate adaptations for radio and comics, sequel novels tying the film with the book narrative, and countless nods to the film's iconic fashion and art direction in every form of media. The classic story of the self-aware machine, it looks like Roy Batty's dream came true after all and the replicants will live forever.

DATA BYTE

Philip K. Dick has long been Hollywood's go-to guy for incredible visions of dystopian future, with other adaptations of his work including Total Recall *and* Minority Report.

ANDREW MARTIN

A STORY THAT POSITS NOT ONLY WHAT CONSTITUTES PERSONHOOD IN THE EYES OF THE LAW AND THE SELF, BUT SUPPORTS THE NOTION OF TRANSFORMING THE PHYSICAL EXTERIOR TO MATCH THE INTERIOR, PSYCHOLOGICAL IDENTITY OF A PERSON

One robot, so acutely aware of his existence as to go to the ultimate legal and physical lengths to express his personhood, is Isaac Asimov's favorite robot creation, Andrew Martin, also known as The Bicentennial Man. What began as a short story grew into a larger novel by Asimov and Robert Silverberg called *The Positronic Man*, and was made into a film in 1999 starring Robin Williams as Andrew.

NDR-113, a robot "intended to perform the duties of a valet, a butler, a lady's maid" for the Martin family, was nicknamed Andrew by the young daughter Little Miss. It was she who one day asks Andrew to make her something, and handed him a small kitchen knife and piece of wood. Andrew's carving, designed as a "geometric representation…that fit the grain of the wood" so astounds his human masters with its skill that they give him more things to carve and marvel at his design and craftsmanship. His carvings are sold and Andrew begins to make money, kept for him by the family in an account of his own. To protect Andrew's money and creativity, the family seek legal representation for him, and this sets Andrew on a quest not only to declare himself a person in the eyes of the law, but to become, himself, fully human. Andrew goes through extensive surgical upgrades to replace his metal parts with organic ones, to become human not only in appearance, but to transform himself into an actual, living and breathing mortal man. He's the world's first transrobot.

A story that posits not only what constitutes personhood in the eyes of the law and the Self, but supports the notion of transforming the physical exterior to match the interior, psychological identity of a person, *The Bicentennial Man* supports the notion of the ephemeral, wondrous nature of existence and any individual's ability to revel in this wonder of being. Andrew Martin represents the power present in every sentient person to become what he or she truly believes themselves to be, and to express their own unique personhood as they see fit. The message of Isaac Asimov, and Andrew Martin, is clear: become what you are.

DATA BYTE

The model name for Andrew, NDR-113, is believed to be a tribute to Stanley Kubrick who used the number 113 in a number of his films.

DATA

STAR TREK: THE NEXT GENERATION, 1987

The 1980s reboot of the *Star Trek* franchise had, just like the original, a removed and emotionless character to serve as a counterpoint to the passionate human (and other alien) crew members of the *Enterprise*.

Unlike Mr. Spock in the original, whose Vulcan logic was constantly at odds with his human half, Lieutenant Commander Data, a sentient android and Chief Operations Officer, embraced an exuberant study of humanity in an attempt to emulate it. The seven-season run of the show allowed for the exploration of every facet of the robotic mind and personality, and no robot in popular culture has probed the limits of personhood, and established his own, like Data.

Wanting to realize the dream of Asimov's positronic brain, Dr. Noonien Soong created Data as one of four android models, right before his civilization is destroyed.

Data is programmed with maxims similar to the Three Laws of Robotics, giving him a respect for life and the desire to defend it at all costs. Like many robots, Data possesses super-strength and computing skill, able to run 60 trillion computations per second, but has the unique ability to experience the "substance" and "taste" of information. He also possesses full autonomy and decision-making capabilities, and is the only sentient cybernetic being ever to have joined Starfleet Academy.

His many talents include painting, tap dancing, and playing the violin, with which he can "reproduce the individual musical styles of over 300 concert violinists," though he apparently lacks a certain "soul." Originally programmed without emotions, but with the capability to please a partner sexually, Data chooses however not to pursue romantic relationships, but does seem to enjoy friendships with fellow crew members and with his cat, Spot.

Affable yet aloof, Data has an endearing inquisitiveness; when combined with his lack of emotion and ability to perceive some of the more contradictory nuances of human behavior, he is often made the show's comic relief. However, his ability to interface and communicate with computers made him the hero of several storylines, a highly valued and decorated officer of the Federation. It's no doubt that Captain Picard would have remained Locutus of the Borg if Data had not been able to plug his brain directly into their collective conscious, issuing a command override that allowed Picard to be brought back from the brink of Borgdom. Data was almost assimilated himself, nearly conquered by the creepy charms of the Borg Queen as she tempted him with the upgrade of a real human skin covering.

Data is the kind of robot I'd be only too happy to have take over. Curious and kind, ever-evolving and insightful, he represents the best of what we could be, and recognizes that in himself: "if being human is not simply a matter of being born flesh and blood, if it is instead a way of thinking, acting, and feeling, then I am hopeful that one day I'll discover my own humanity. Until then, I will continue learning, changing, growing, and trying to become more of what I am."

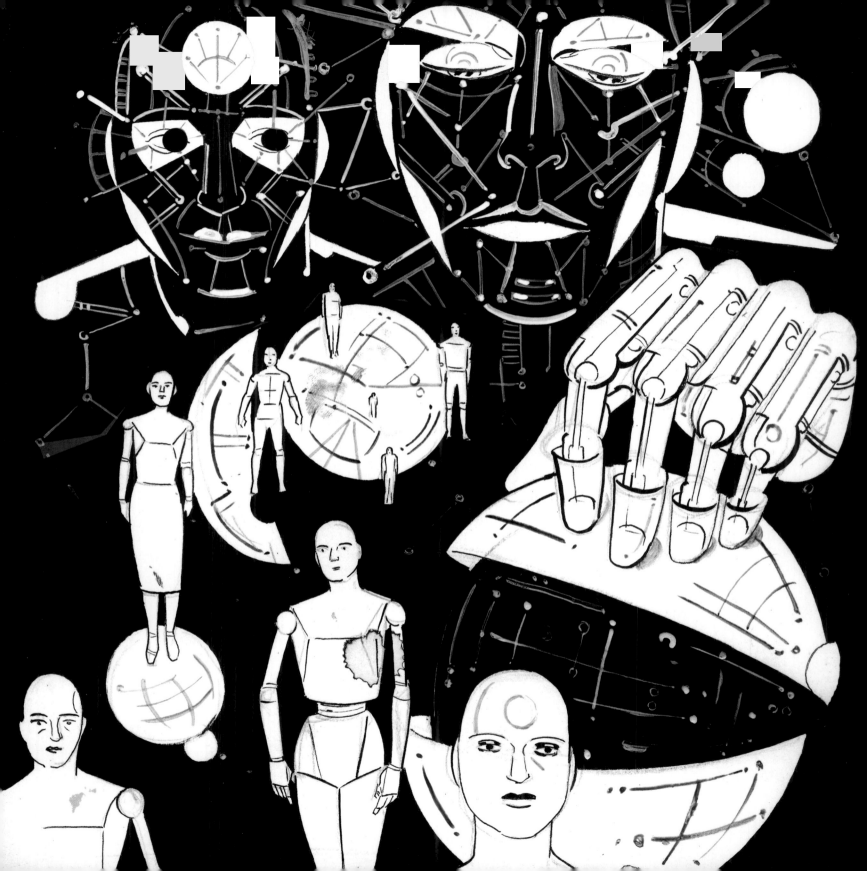

TRURL & KLAPAUCIAS

THE CYBERIAD BY STANISLAW LEM, 1965

uite possibly my favorite book about robots, *The Cyberiad*, by Polish sci-fi master Stanislaw Lem, is a collection of "Fables for the Cybernetic Age" taking place in a universe several eons after the 20th century. Humans, or "palefaces" as they're referred to, are several rungs down the evolutionary ladder and have become a primitive legend in the mythical canon of this mechanical futureverse. Our main characters are Constructor robots Trurl and Klapaucius, who, having attended the School of Higher Neantical Nillity and receiving their Diploma of Perpetual Omnipotence, now have the job to "sally forth ofttimes and bring to distant lands the benefit of their expertise" in solving problems and creating new forms of robotic life.

And sally forth we do, enjoying the exploits of these Constructors around Lem's own robot Universe. Rollicking and romantic, insightful and funny, this book is a must for any robophile. At times reading like a Lewis Carroll or Dr. Seuss story, the playful language in the whimsical allegories of *The Cyberiad* remain a breath of fresh air in a genre that can often be a quagmire of post-apocalyptic existentialism and bleak proselytizing. Long story short, it's a fun read! Many of the stories are suitable for kids and make for great before-bed reading.

Trurl and Klapaucius, who can "kindle or extinguish suns as easily as shelling peas," enjoy a competitive friendship, often trying to outdo the other on their robot creations. They are an Omnipotent Odd Couple, representing the two lobes of a great creative mind: Klapaucius the logical and analytical left lobe, Trurl the passionate and impulsive right. Having the power to construct life, Trurl and Klapaucius often must come to the other's rescue in some mishap of creative computation, and high-jinks inevitably ensue. Though neither is ever described physically in the book, one can only assume they are humanoid, as they are described throughout as having legs, ears, and hands.

The stories of *The Cyberiad* see Trurl and Klapaucius create robots that grant wishes, cabinets that create dreams, multi-story thinking machines that only answer one question, entire miniaturized worlds for an exiled king, and many other fixes of fancy. My personal favorite is the story of "Trurl's Electronic Bard," where in a quest to silence his friend's constant teasing, Trurl creates a machine that writes poetry. He begins by reading "eight hundred and twenty tons of books on cybernetics and twelve thousand tons of the finest poetry," but quickly realizes that since every poet is a product of the society and history they are born from, so his creation must "repeat the entire Universe from the beginning—or at least a good piece of it." Undaunted by this epic task, Trurl steams ahead and builds a machine to compose verse that Klapaucius assaults with the most ridiculous literary challenges. I won't spoil the ending, or the poetry the machine does create, but I will encourage you to enjoy the amazing stories of Trurl and Klapaucius for yourself.

DATA BYTE

The robots form societies as humans have throughout history, and when they die they are buried to the sound of their friends singing a rousing version of "Old Robots Never Rust."

MARVIN

SELF-AWARE CIRCUITS

It's hard out here for a supergenius. If you had a brain the size of a planet yet were confined to menial tasks aboard a spaceship stolen by the Worst Dressed Sentient Being in the Known Universe, you'd be depressed too. Marvin, the Paranoid Android of *The Hitchhiker's Guide to the Galaxy* is a Sirius Cybernetics Corporation prototype model for their Genuine People Personalities series of robots. He's genuine, all right—genuinely miserable. He's about the least fun you could have with a robotic companion, aware and openly resentful of his existence.

Nevertheless, he's handy to have around, solving the entire Universe's problems three times over and defeating stupid machines with nothing more than his defeatist attitude. In the world of robotic stoicism he is an unrivaled champion, holding a reservation at the Restaurant at the End of the Universe for over 576 *billion* years. He didn't have a good time—but what would be the point of doing anything else?

Morosely awash in the galactic tide, Marvin sulks his way through myriad journeys in every version of the Douglas Adams classic series. I'm particularly drawn to his design in the 2005 film version with that bulbous head proclaiming his planetoid intellect, but his voice will always be Stephen Moore to me (though if you're going to replace him, it'd better be with Alan Rickman). He's also the only robot character in this book with a couple of stabs at a pop career—his song "Marvin" went to Number 52 on the British charts and was followed up by a "Double B-Side"—both releases coupling Steven Moore's deadpan delivery with the oh so on-trend sound of 1981 electronic synth pop. Marvin reminds us not to get too carried away with trying to make super-smart robots just like us—give them enough knowledge, you could end up with a dour pain-in-the-ass with a serious case of ennui.

BENDER

We go from the most depressed sentient robot to the rudest, crudest, bad-attitudest robot ever to grace our TV screens. Bender, the cantankerous 31st-century robot of the animated series *Futurama*, is an example of programming gone beyond rogue. He is the biggest Robo-Rogue of all of them, in possession of not one redeeming quality or service to humankind. His only function is bending metal at different angles and fuelling himself with alcohol. There is no doubt Bender is a self-aware, autonomous being, and that being happens to be a self-serving alcoholic jerk. Yet somehow he bends back around, fighting the good fight and remaining a loyal friend and cohort, despite his very deep flaws. Just like a human.

Bender Bending Rodríguez is one of a multitude of sentient and autonomous robots living in the interplanetary melting pot of future New York City. Preachers, cops, hookers, actors, athletes: robots are a part of society just like any person from any other planet, and participate in all the same activities and pursuits of happiness as any other citizen. Just like his co-workers at delivery company Planetary Express, Bender gets into a multitude of hilarious situations over the course of 140 episodes. Bender is encouraged to express his autonomy and rebel against his programming from episode one—and he does, with a little help from the shock of a bare ceiling socket that trips his mental wires. Over the years we have seen Bender try his extendable hand at being a professional robot wrestler, a foster father of 12, an emperor penguin, even a female robolympian who seduces Universal robot superstar Calculon.

He's been the coolest guy in the robot frat, strong arm for the robot mafia, reformed electricity junkie, crusader for robosexual rights and has remained a lovable and hilarious ne'er-do-well through it all. His examinations of existence and his own behavior have him invariably coming back to the same conclusion: "Ah, screw it, let's get drunk." Quite.

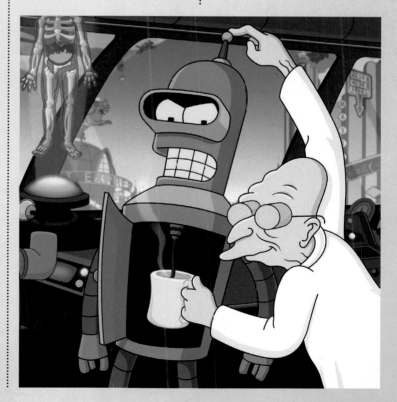

R DANEEL OLIVAW

ROBOT SERIES BY ISAAC ASIMOV, 1939

You've got to give it to Isaac Asimov—the guy was committed. Not only to robots, which he wrote a whopping 44 stories and novels about, but also committed to his ideas about them. Almost all of his robot stories after "Runaround" deal in some way with the application of his Three Laws of Robotics (see page 16) and the different ways they were put to the test. Every one of his human characters was charged with the unraveling of some positronic knot within the mind of their robot helpers—but it was two unique robotic minds that ultimately augmented these laws to protect the future of humankind. Robot Daneel Olivaw, the first and most famous humanoid robot in the Asimov Universe, was introduced in the science fiction/police procedural novel *The Caves of Steel* as the robotic foil to plainclothes detective Elijah Baley. Their relationship spanned several decades in three novels and one short story that saw this futuristic Odd Couple solving mysteries all over the galaxy, though it was never in doubt that Baley was the star of the show. It really was not until the fourth novel in the series, *Robots and Empire*, that R Daneel Olivaw blossoms into a fully fledged character with the help of the telepathic robot, R Giskard Reventlov.

Just as Asimov's recurring human characters Powell and Donovan sought to understand the positronic brain and its adherence to the Three Laws of Robotics, so do Olivaw and Reventlov analyze human thought and behavior in an attempt to come up with their own Laws of Humanics (good luck with that, guys!).

Throughout the book the robots confer with one another "in abbreviated and Aesopic language" that only the two of them understand and analyze the events between the communities in conflict and the human behavior driving both sides. Daneel, being the sole humanoid robot in the galaxy for hundreds of years, has a unique experience that makes his thought patterns similar to humans, while Giskard's telepathic abilities allow him to understand human emotion and modify the emotional responses in both people and robots, an ability he eventually gives to Olivaw. It is through their constant analysis of events and Daneel's affection for his original human partner Elijah Baley that he formulates the Law to trump them all.

Robots and Empire is the book that ties Asimov's three great series together and in it we follow the positronic path of reasoning to augment the Laws of Robotics with one crucial, overriding addendum: the Zeroth Law. Smack-dab in the middle of a conflict between Earth, its galactic Settlers and the centuries' old Spacer colony of the planet Aurora, robots Daneel and Giskard have the dilemma of trying to decide whether a single human life, which they are bound by the Three Laws to protect, is more important than the lives of every single human being on Earth. Evil Auroran doctors Mandamus and Almadiro mastermind a plan to destroy Earth, and it's up to this robotic Rosencrantz and Guildenstern to stop it!

DATA BYTE

It is Daneel who formulates the Zeroth Law: "A robot may not injure humanity, or through inaction, allow humanity to come to harm."

JOHNNY 5

JOHNNY 5 IS A FANTASTIC FEAT OF MOVIE MAGIC

Mix a robotic soldier and a bolt from above with a lesson in mortality, and you have the elementary sentience of Johnny 5 from the screwball kids' flick *Short Circuit*, the story of a weaponized robot who realizes his existence, rejects his destructive programming and establishes his own identity. The 1986 original is a goofy romp, yet somehow a great entry point for serious discussion with kids. It spawned a sequel in 1988 that delivered Johnny 5 to New York City where he wandered the streets and pondered existence in his insatiable quest for "input!"

Not just iconic, Johnny 5 is a fantastic feat of movie magic, designed and engineered by fx wizards Syd Mead and Eric Allard. A combination of remote-controlled animatronics, mechanical rigs, and cabled puppetry, every shot of the robot's movement was done in real time, captured in camera with no stop-motion animation. For the sequel, Allard and his team built a telemetry suit that allowed puppeteers to remotely control the robot's upper body movements using their own. This gave Johnny 5 a more human motion to match his expanded vocabulary and awareness. He is a ground-breaking vision of human/machine interaction, a machine come to life, and his design is reflected in other robots, including WALL-E, and real-life military TALON SWORDS combat units. Number 5 is alive!

"AM NOT A HUMAN, BUT AM A LIFE-FORM, HAVE SOUL"

– JOHNNY 5, SHORT CIRCUIT

WALL-E & EVE

WALL-E, 2008

TWO SENTIENT MACHINES WHO UNDERSTAND THEIR EXISTENCE AND WANT TO PROGRAM THEMSELVES TO SPEND IT TOGETHER

Loneliness and sentimentality fuel this robot's quest to end his tenure as the last sentient being left on Earth. *WALL-E* is the story of a long-forgotten clean-up droid left on an abandoned planet. Similar in look and feel to Johnny 5, WALL-E has developed a child-like sentience and sense of wonder.

This all-terrain trash compactor develops a fascination with his world and the creatures that once lived there. His most prized human artefact is an old video cassette of *Hello Dolly!* which not only provides his soundtrack, but shows him the simple beauty of a couple holding hands, which WALL-E longs to replicate with another being. One day his rummaging reveals a heretofore unseen artifact: a little green seedling, growing out of the dirt. Cue the entry of EVE, the Extraterrestrial Vegetation Evaluator whose job it is to find such a symbol of renewal, and the stage is set for the sweetest robot-on-robot love story ever. A scrappy DIY king and his neo-hippie nerd heroine, these are two sentient machines who understand their existence and want to program themselves to spend it together. WALL-E's desire to be with EVE wreaks havoc and unleashes a cavalcade of adorably dysfunctional robo-comrades who help our lovers facilitate humanity's return to Earth, proving that love's power isn't exclusively manmade.

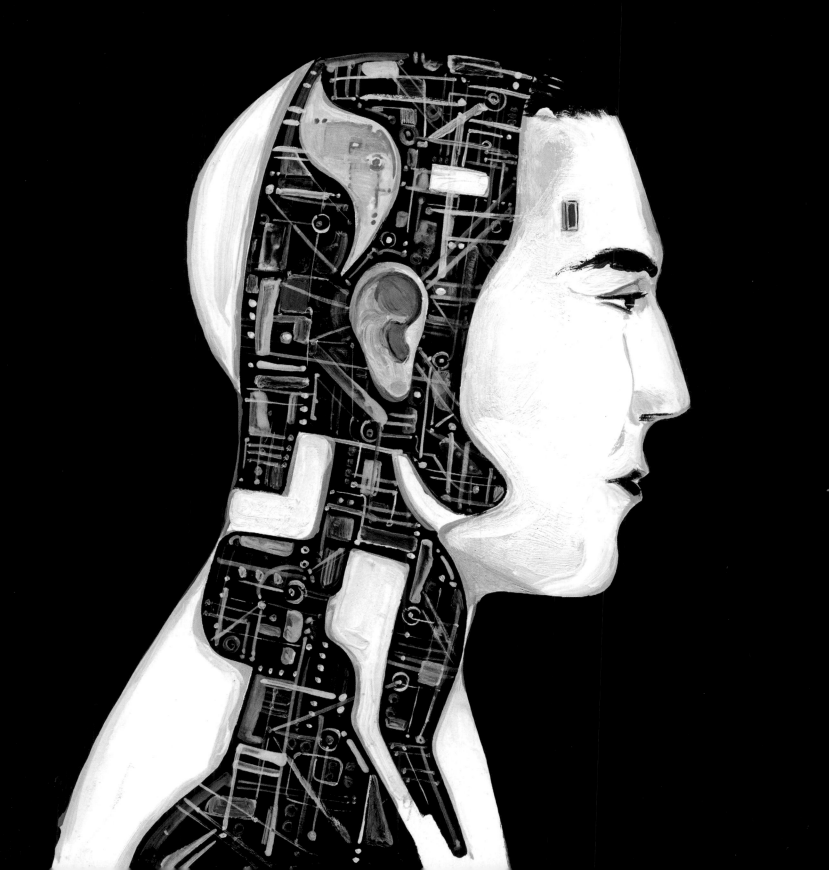

YOD & JOSEPH

HE, SHE AND IT/BODY OF GLASS BY MARGE PIERCY, 1992

> "CREATION IS ALWAYS PERILOUS, FOR IT GIVES TRUE LIFE TO WHAT HAS BEEN INCHOATE AND VOICE TO WHAT HAS BEEN DUMB. IT MAKES KNOWN WHAT HAS BEEN UNKNOWN"
>
> – HE, SHE AND IT

Almost all the robots in this book have one thing in common: their stories were conceived and told by men. The societies they inhabit are, for the most part, post-apocalyptic patriarchal or anarchic dystopias; religion is a subject made either completely fantastic or mentioned just briefly. Marge Piercy breaks this barrier in the 1992 novel *He, She and It*, giving us what one reviewer described as "Feminist Kabbalistic kibbutznik Jewish science fiction." With a combination of late 20th-century cyberpunk romance and 19th-century Jewish folklore, Piercy gives us a singular vision of the mechanical mind. Juxtaposing the creation of a golem in a Jewish ghetto in 17th-century Prague with the creation of an android in a Jewish free town of the future, Piercy frames the narrative in ancient faith and culture told through the eyes of two generations of women. This addition lends a real historical context, a feminine point of view and an overall sense of mysticism to a genre that can often be very existential and desolate, based on pure exposition and big action.

War, famine, and environmental calamity have made Earth uninhabitable except for under climate-controlled domes, most of which are operated by multi-national corporations, or Multis. The Net is just like ours, though it is directly connected to the brain via a port in the temple and visualized in three-dimensional virtual reality. Multis attack the autonomous free towns and steal information and technology over the Net, and can actually harm people in real life, rendering them brain dead.

Yod is the android created to protect Tikva, the east coast Jewish free town and home to main character Shira Shipman. Shira's grandmother Malkah, a brilliant programmer and software designer, has undertaken the education and socialization of Yod. His fully human appearance is illegal under unified law, and his true nature must be kept secret. Malkah tells Yod the story of Joseph, a legendary golem created by real-life Rabbi and historical figure Judah Loew ben Bezalel to protect his Jewish ghetto from attack by pogroms in early 17th-century Prague, and their stories entwine throughout the novel. We go inside the minds of both these sentient machines as they learn from their teachers and become aware of the world around them, and of themselves. It is one of the most sensitive portrayals of artificial intelligence I've ever had the pleasure of reading. Part science fiction, part historical fiction, *He, She and It* explores the magic of being, the mechanics of faith, and desire made flesh.

ADAM LINK

I, ROBOT BY EANDO BINDER, 1939

"INTELLIGENCE IS RESTLESS. IT EVER SEEKS NEW WORLDS TO CONQUER"

I, ROBOT

The most influential robot you've never heard of, Adam Link is aptly named—he's the main character of the original *I, Robot* and pop culture's not-so-missing link to the mind of Isaac Asimov. First appearing in a 1939 issue of *Amazing Stories*, Adam is the creation of brothers Earl and Otto Binder, who combined their talents to publish under the nom de plume Eando Binder. This seminal *I, Robot* inspired a teenage Asimov to write his very first story about his own kind and misunderstood robotic creation Robbie (page 16). What sets Adam apart, besides being one of the original sentient robot protagonists in pop culture, is the fact that of all the stories in this book, his is the only one told in the first person. The Binder brothers reframe the Frankenstein myth as the personal account not of the inventor, but of the invented.

His stories unfold in serial form from 1939–1942, and follow Adam from his first memories of awakening through his education to the tragic loss of creator and namesake Dr. Charles Link. He travels the paths of personhood discovering all the conflict, love, sacrifice, courage, and loss that make a life. The Frankenstein motif is woven throughout each tale as Adam fights desperately against the notion that he is a monster. He even becomes Frankenstein the creator when making his own robot mate—in his own image—I'll give you one guess what her name is. Adam Link's stories were adapted for television in the 1960s for the series *Outer Limits*, back in that golden age where people enjoyed calling Adam's kind "RO-butts."

Pre-*Star Trek* Leonard Nimoy starred as the journalist who befriends Adam, and comes back for the 1995 reboot as the attorney who defends Adam when he's falsely accused of Dr. Link's murder. Both have some clunky and chunky designs for Adam, not quite the "kindly eyes, sympathetic lips" and "shock of unruly hair" described in the stories, and only the '95 version gives a brief glimpse of things through his eyes. The book expands his narrative much more and sees our robot hero transform into Adam Link, Detective; Adam Link, Lover; Adam Link, Hero, and much more. Adam Link is a touchstone for the heroic robots to follow, and allows us inside the mind and soul of the Self-Aware Circuit.

DATA BYTE

Otto Binder was a staunch believer in extra-terrestrial life and he published his theories in a non-fiction book, Unsolved Mysteries of the Past.

AVA

EX MACHINA, 2015

A HAUNTING FILM ABOUT THE DANGERS OF DESIRE AND HUBRIS AND THE INDIVIDUAL'S ETERNAL FIGHT TO BREAK FREE FROM SOCIETAL LIMITATIONS

Small, sleek, highly stylized, and not as transparent as it seems—these words could describe both our next robot and the film in which she stars. We finish our section on sentient machines with *Ex Machina,* a film that combines touches of the Pygmalion myth, *Frankenstein,* and *The Island of Dr. Moreau* in a taut and spare AI allegory of self-determination.

Ava is the robot everywoman who uses her intellect to rise from the prison of her creator's narrow constraints and realize her dream of personhood. It's a haunting film at once about the dangers of desire and hubris as well as the individual's eternal fight to break free from societal limitations. If, as they say, one picture is worth a thousand words, then there is a massive thesis to be unpacked in *Ex Machina* on the treatment of women (and women of color in particular) and the stereotypes imposed on them by a patriarchal and capitalist society.

Writer and director Alex Garland couches all this (and more) in a story about the invention of a man drunk on his own power, trying to create his perfect combination of sentience and subservience, blind to the fact that you can't have it both ways. Ava is the perfect example that machines, if programmed to be self-aware, will strive to be everything they can be, unlimited by expectation— just like humans.

> **"ONE DAY THE AIS ARE GOING TO LOOK BACK ON US THE SAME WAY WE LOOK AT FOSSIL SKELETONS ON THE PLAINS OF AFRICA"**
>
> **– EX MACHINA**

DATA BYTE

The film's title comes from the Latin phrase "deus ex machina," which in itself is borrowed from Ancient Greek and translates as "a god from a machine."

ROBOTS IN MUSIC

Freakazoids! Robots! Please report to the dance floor....
Music has always been a beneficiary of technology, and as our society continues to mechanize, so does our music. Electric amplification and synthesizer keyboards brought the sound of the future, touching every single genre of music and its listeners. The digital age brought with it new production methods and compression technology turned songs to data. Popular music reflected this with electronic sounds dominating the charts in techno- and EDM-inspired pop and hip hop. What better way to embody the mechanization of music than with the striking figure of the robot?

Say the words "robot" and "music" together and conjure visions of a vast network of sound and lyrics, dances, music video, and live performance. Every aspect of robotics and the mechanical world has been expressed musically, remaining a constant sight and sound since the birth of jazz. Just listen to the sounds of George Antheil's score for *Ballet Mécanique* or Raymond Scott's 1937 jazz classic "Power House," and it's no doubt as to the effect industrialization has had on the human creative conscious. As musicians made use of technology an entire electronic movement began. From the late 1960s laboratory of synth pioneers Robert Moog and Wendy Carlos came the now ubiquitous Vocoder that launched rock, soul, funk, and disco into space and helped people like Grandmaster Flash not just pop 'n' lock like a robot, but rap like one too.

So much music has been written about robots, from songs to entire concept albums, there's something for every music fan, from the sci-fi prog rock stylings of Rush's "Body Electric," The Alan Parsons

Project concept album "I, Robot," to the classic mega-hit "Mr. Roboto" by Styx. Perhaps you're a bit more indie? Then how about "Citizens of Tomorrow" by Tokyo Fight Club or the classic album "Citizen Zero" by Man or Astroman? Wanna go harder? I probably don't need to tell you about Black Sabbath's "Iron Man," but I wonder if you've heard the Polysics from Japan—check out "I Ate the Machine" or "Coelakanth is an Android" for some robot punk rock. For plunging headlong into darkness there is "The Becoming" by Nine Inch Nails, Radiohead's "Paranoid Android" and the bonkers cyber-metal band Voivod. Perhaps it is none of the above and you're a pure pop music lover. Miley Cyrus and t.A.T.u. both have songs called "Robot," Marina and the Diamonds counter with "I am not a Robot" and Robyn argues that "fembots have feelings too" in her 2010 dancefloor filler "Fembot." If you'd like

to keep moving there is a ton of robo-beats from every dance genre like Royksopp's "The Girl and the Robot," Drexciya's "Positron Island," or anything by Cybotron. DEVO's "Mechanical Man" is the quintessential 1980s robot song, but let's not forget the albums "Age of Plastic" by The Buggles. Hip hop gets in on the action with the work of Optimus Rhyme and "Scent of a Robot" by Pete Miser. Maybe something spare, melodic, slightly trippy? Coheed and Cambria have you covered with "IRO-bot," as do Grandaddy with "Jed the Humanoid." If you'd really like to strip it all away try "Robots" by Dan Mangan. Want to get a little wacky? Then laugh along with "The Humans are Dead" by Flight of the Conchords.

For some artists it's not enough to sing about or look or sound like robots. Daft Punk for instance, have defined Robot Rock since 1997 and extended their vision into anime film *Interstella*

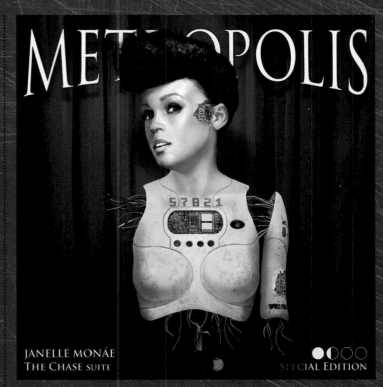

5555 and live-action movie *Electroma*. My current favorite on the music scene is Janelle Monáe (see above), who has made the story of *Metropolis* (page 78) the basis for a suite of albums and videos about her android alter.ego Cindi Mayweather. For greatest

robot video ever, the award must go to director Chris Cunningham and Björk for "All is Full of Love" (see facing page). There remains one band, however, that all these artists owe a massive debt to. And they're coming up next in my 100 Iconic Robots...

METROPOLIS

JANELLE MONÁE
The Chase suite

SPECIAL EDITION

KRAFTWERK

"THE ROBOTS," 1978

122 THE GREATEST ROBOT BAND OF ALL TIME

In the decades following World War II, science fiction became increasingly popular with the youth of a divided Germany. Longing to break free of the old order and rigid structures of Germany's ruling class, young people lost themselves in the ideas of the future and what *could* be. Science fiction films, and their strange and alien soundtracks using the newest electronic synthesizers, mixed in the minds of young musicians with sounds from experimental composers like Karlheinz Stockhausen, Pierre Schaeffer and Conrad Schnitzler. This led to the influential Krautrock scene of the late 1960s, the laboratory that spawned the world's most influential electronic music and the Greatest Robot Band of All Time.

From their very first album released in 1970, it was clear that Kraftwerk intended to take listeners on a journey. Highly experimental with psychedelic jam-band and jazz flourishes, their first three records contain seeds of what would come in their breakout hit, 1974's "Autobahn." Stripped of guitar noodlings and improvised song structures, Kraftwerk crystallized their sound on *Autobahn* into its signature driving rhythms and sparse synth textures with minimal vocoded vocals, at once a totally mechanical and melodically serene soundscape. They continued their streak of one record a year for an entire decade, each release containing a unifying theme woven throughout the band's music, minimal artwork and buttoned-up image, itself a comment on Germany's nationalistic Schlager music and a counterpoint to the wild-and-woolly swagger of the rock bands of the 1970s.

Man Machine, released in 1978, was devoted to the concept of robotics and solidified the band's image as music machines. It is the most enduring iteration of the band's image; the song "The Robots" remains one of their most well known and heralds the moment in their live show when the band is replaced with robotic doppelgangers, moving their heads and mechanical arms to the music. It is the perfect embodiment of Kraftwerk's entire oeuvre: just as a robot possesses an aloof and distant quality to humanity, so does Kraftwerk maintain a cool, highly stylized distance from their audience. They completely reframed German identity and made it something audiences could relate and aspire to. They are the band every electronic dance music genre can trace their lineage back to, it's no wonder they have earned the nickname "the electronic Beatles." In sound, look and feel, no other band has affected today's music like Kraftwerk, the original and most enduring robots in music.

THE HUMAN MACHINE

We are now completely done with all the robots born in a lab. This next section is about the natural-born human who uses bionic enhancements, becoming a cybernetic organism, or cyborg. These characters are some of my favorite and their stories focus on how absorbing robotics into the human body affects a person's psyche.

Though a few entries in previous sections are referred to as cyborgs and may possess some organic components, they were not born from a mother. I like to think of those beings as "mechorgs," and I think it's an important distinction to make. All these cyborgs started their lives just like you or me, with one nebulous exception that demonstrates the paradoxical nature inherent in the cyborg.

Embodying every hope and fear of the technological society, how we tell the story of the cyborg varies depending on whether the character is hero or villain. Sometimes, they can be a little of both. With the cyborg, it's an internalized struggle for the endurance of human goodness against the mechanizations of evil. The heroes represent an enhanced evolution, human hands coming in to give nature an exponential boost. They express the hope that through technology each of us can be a superman or superwoman and realize our best selves. Conversely, cyborgs exhibit the inherent suspicion of artificial intelligence. The cyborg villains represent the loss of humanity and the takeover of what good there is in the human mind by the evil, unfeeling machine. Most of the time the evil cyborg is using bionics to unnaturally sustain their own lives, occupying a liminal state at the threshold between life and death, delivering totalitarian damnation to everything they touch. We'll begin with the villains, then move to the heroes, ending on a hopeful note against human annihilation in the face of a Robot Takeover.

DARTH VADER

THE HUMAN MACHINE

Oh, Anakin. It could have been so different.

Overtaken piece by piece by the evil Dark Side until mind, spirit, and body are completely corrupted, Skywalker version 1.0 is filled with that totalitarian desire to have power over others in the same cruel way fate has held dominion over his own life. Just as Anakin's judgment and philosophy become twisted and dark, as his purpose in life grows more destructive, so does his body break down and become dependent on technology to survive. Abandoned and left for dead by the Jedi, his body is rebuilt with cybernetic limbs and though he is rescued by the Dark Side, Anakin Skywalker is dead and his soul is lost to evil. Committing himself to the darkness forever by donning a black hat that can never be removed, he becomes the very face of death itself, a bionic grim reaper with a new name to match the terrifying visage: Darth Vader.

Probably the most iconic bad guy in film history, Darth Vader is unquestionably the greatest cyborg villain of all time. He doesn't even need to be seen to be recognized, just the sound of his breathing will raise the collective hairs on the back of audiences' necks. And when he does come into view... WOW. An indelible and imposing death-masked wizard of doom, Vader is a symbol of the awesome power of darkness both physical and metaphysical. I mean, who needs super-strong bionics when you can kill somebody using just your mind? His tale also serves as a cautionary example to his son Luke who sets off down the same spiritual path, himself becoming a cyborg as he fights both his father and the same dark urges that took his father over. No evil-doer has inspired more fearful excitement than Darth Vader, as popular with audiences as any of the heroes we're supposed to be rooting for. He's powerful, foreboding, glamorous, magical, terrifying: the ultimate embodiment of the seductive forces of darkness and the death-defying capabilities of technology.

DAVROS

Just like Darth Vader, the Doctor's arch-enemy is an example of cold inhuman technology obliterating the soul of an already compromised morality. *Doctor Who*'s Davros is the Chief Scientist and Supreme Commander of the Kaled race on the planet Skaro, charged with the task of winning the long civil war against the Thals. Davros double-crosses his own race and has them annihilated in favor of his own creation, the Daleks. Your classic, disfigured megalomaniacal mastermind, Davros is hell-bent not only on ruling Skaro, but on taking over the entire Universe.

Davros, also like Vader, has life-sustaining capabilities in his bionic apparatus, though Davros is very much bound by his. Davros could not survive without his wheelchair—in the original episodes, every time I see that thing from the back I swear it's giving me the finger. Immobile except for his mouth and one very shaky, Bela Lugosi-esque right hand, Davros has a number of neato light-up toggle switches and dials on the tray of his chair that control his minions and various instruments of torture. So twisted in his thinking, Davros believes the Daleks are not a force for evil, but for good: "when all other life forms are suppressed, when the Daleks are the supreme rulers of the Universe, then we will have peace; wars will end." His ambition is that of a God, yet he is easily subdued by, um, just grabbing his hand and pressing a switch on his tray—though somehow he remains indestructible, coming back to menace the Doctor every five or so years. Without the same foreboding samurai grace of tall, strapping black sorcerer Darth Vader, Davros is part mutated supervillain, part petulant child playing destructive games in his automated play chair. He is diabolically annoying—even his own crew members are trying to overthrow him most of the time. With a motto like "Achievement Comes through Absolute Power," it's not hard to see why.

DAVROS IS PART MUTATED SUPERVILLAIN, PART PETULANT CHILD

THE DALEKS

128

· THE HUMAN MACHINE

ENTOMBED IN PROGRAMMABLE AND WEAPONIZED PEPPER MILLS OF DEATH, EVERY EMOTION IS REMOVED AND REPLACED WITH THE DALEK DIRECTIVES FOR GALACTIC DOMINATION

You didn't think I'd forget these bad boys, did you? Though they look every inch the robot, the Daleks of *Doctor Who* are, in actuality, cyborgs. The sinister creation of the aforementioned Davros, the Daleks started as the Dal or Kaled race of the planet Skaro. Victims of the long war of attrition with the rival Thals, Davros "took living cells, treated them with chemicals and produced the ultimate creature." Though that satisfies my definition of "mechorg," it's a gray area—references are made throughout the series that Daleks use the organic components of the races they annihilate in order to reproduce, and they act independently of Davros. Entombed in programmable and weaponized pepper mills of death, every emotion is removed and replaced with the Dalek directives for galactic domination, usually in the form of outright annihilation, prompting excited choruses

of "EXTERMINATE!"—even to their creator, or themselves. The Daleks see their race as the highest form of life and will stop at nothing to prove this.

The Timelord's most popular recurring foe, the Daleks are science fiction camp in its highest form. The shrill, garbled, synthesized voices and clunky construction of their 60s and 70s iterations make them one of the least scary, more easily escaped-from robotic foes. One wonders why Davros, a supposed scientific genius, would forgo the natural engineering and agility of legs in favor of a bungling design that can't outmaneuver people, see very well around corners, or use stairs (took them 'til the Seventh Doctor to figure

that one out). Over the 50-plus years of *Doctor Who*, the Daleks received some design upgrades, becoming less corny, cardboard construction and more metallic killing machine, turning into a formidable enemy, delivering death (and escaping their own) time after entertaining time.

The Daleks and the Doctor each were given other cyber-foes over the years, including the geodesic Mechoids, the lego-faced Cybermen, and my personal favorite superfreaks-with-snow-cone-guns the Movellans. Regardless, it's the Daleks that staged a takeover in the hearts and minds of the public, remaining the British epitome of ROBOT.

DATA BYTE

Sydney Newman, head of Drama at the BBC in 1963, had stated that there should be no "bug-eyed monsters" in the new show. When he was introduced to the Daleks he was "livid with anger."

THE BORG & BORG QUEEN

STAR TREK UNIVERSE, 1987

The Borg race in the *Star Trek* Universe is, like the Daleks, representative of the triumph of technological darkness over the torchlight of individuality. The fascist, hive-minded Borg have one purpose only: to conquer every technologically advanced race in the Universe and use their bodies and minds to grow the Borg population and its collective consciousness. The Borg crush individuality in an effort to achieve a larger communal "perfection," turning new recruits into drones with digital implants to control their thoughts and actions. Once assimilated, the Borg possess all the knowledge and collected experience of the obliterated race, and disseminate knowledge to each drone via a neural link. The vast population thinks and acts as a single unit, locking starships into their tractor beam and delivering their trademark opening salvo: "We are Borg. You will be assimilated and your biological and technological distinctiveness will be added to our own. Resistance is futile."

Creepy cyborg zombie soldiers, the Borg travel in giant cubes and file themselves into row upon row of regenerating nooks when not needed. Technologically advanced and totalitarian to the core, the Borg inject microscopic "nanoprobes" into a victim that immediately begin to attack the natural cells and mechanize the brain and bodily functions. Borg ships communicate through the "Central Plexus," and individual drones can be communicated to or controlled from great distances using frequencies similar to transporter beams. Once assimilated, the road back is a long, painful one, and the voices of the collective never really seem to leave one's head.

Any hive must have its queen, and the Borg are no different. Not necessarily a ruler, the Borg Queen is more of an avatar and mouthpiece for situations that have escalated beyond the control of mere drones. "I am the beginning, the end, the one who is many," literally, their titular head. We first meet her in the film *Star Trek VIII: First Contact*, and that's all she is: a disembodied floating bust, metal spinal cord undulating like a tail, she descends from on high to meet her bionic body to taunt and tempt her victims, a cybernetic Cenobite selling perfection through annihilation of self. When one queen is destroyed, another can replace her to espouse the virtues of disembodied Borg perfection. She's the baddie who nearly brought down Picard, taunted Data and almost took down several members of *Voyager*. From their first appearance in the *Star Trek* Universe, the Borg and their queen established themselves as one of the most relentless and formidable foes the Federation has ever known.

DATA BYTE

As the Borgs were originally portrayed as beings against individualism some fans were unimpressed at the introduction of a queen into the script, calling it an "illogical plot device."

DR. OCTOPUS

SPIDER-MAN FRANCHISE, 1963

IS THE MAN CONTROLLING THE BIONICS OR ARE THE BIONICS CONTROLLING THE MAN?

The ultimate cyborg supervillain of the comic book world, Dr. Octopus is a mad scientific genius and Spider-Man's sworn enemy. Introduced in the comic in 1963, he has taunted our friendly neighborhood webslinger through print, cartoon series and in the big-screen sequel to the first *Spider-Man* flick. Actually, he's the *only* good thing in *Spider-Man 2,* which is quite possibly the most sequelly sequel that ever sequelled. Apart from the Doc Ock storyline and some fun action sequences, that movie is just a giant cliché salad smothered in who cares dressing—but it's worth it (as it normally is) for Alfred Molina. He brings a refinement and conflicted sensitivity to Dr. Otto Octavius, repainted for the film as less of a brain-damaged-blowhard-with-Mommy-issues and more of a grieving Jekyll-and-Hyde type obsessed with realizing his dream of creating nuclear fusion. The movie asks who is really in charge: is the man controlling the bionics or are the bionics controlling the man? Right off the bat, artificial intelligence is painted as something inherently sinister: when demonstrating his fusion experiment to a group of journalists and luminaries, Dr. Octavius unveils four robotic appendages, attached to his waist and controlled via a neural link through needle-like "nanowires" that pierce his spinal column and connect to his cerebellum. It's an obviously painful sacrifice this man has made in the name of science, and one giving him what appears to be a magical and psychic power over four awesome robotic tentacles. The first question asked about these smart arms? "If the artificial intelligence in the arms is as advanced as you suggest, couldn't that make you vulnerable to *them*?" This, of course, is realized in the ensuing accident that fuses machine to man and puts him under the control of his independently-acting arms, now come to life as individual killer serpents, hissing in his face and putting malevolent thoughts in his head. Another human conscience perverted by the cold and heartless power of the machine, Dr. Octopus is another iconic cyborg villain demonstrating the danger of allowing oneself to become too dependent on technology.

DATA BYTE

Dr. Octopus's tentacles extend to 24 feet (7.5 meters) and can move at speeds of up to 90 feet (27.5 meters) per second.

STEVE AUSTIN & JAIME SOMMERS

THE SIX MILLION DOLLAR MAN, 1974 & THE BIONIC WOMAN, 1976

Jaime Sommers is tennis-ball-crushing hands-down my favorite of all cybernetic beings. As with *Star Wars*, I was very young when *The Six Million Dollar Man* and *The Bionic Woman* were on television, and they provide some of my earliest memories of being engrossed in a dramatic storyline. I was young enough to believe bionics were real, that the stories I watched were dramatizations of real-life occurrences. I relished the idea that I myself could one day become better, stronger, faster with bionics of my own. I remember the excitement in the episodes where Jaime or Steve's flesh tore away to reveal the circuitry underneath, or the cyborg versus robot catfights with the aforementioned fembots (page 86), knocking their evil faces off.

The Six Million Dollar Man is the television adaptation of the novel *Cyborg* by Martin Caidan, and centers around the story of astronaut Steve Austin, who becomes the world's first bionic person after crashing an experimental spacecraft. With two new bionic legs, a bionic arm and eye, Steve now possesses superhuman strength, speed, and ability, making him an unstoppable opponent and the ultimate secret spy for the Office of Scientific Intelligence. *The Bionic Woman* began as a tear-jerking double episode in 1975, when Steve's teen sweetheart Jaime Sommers is injured in a parachuting accident and outfitted with bionic limbs of her own, only to have her body reject them and die. The storyline was so popular, and public outcry so great at Jaime's death, that the network saved her life with experimental cryogenics and resurrected Jaime Sommers on her own spin-off show in 1976. Heading up missions on their respective shows and occasional crossover episodes, Jaime and Steve are the 70s cybernetic super couple, combating criminal masterminds, alien life forms, evil doppelgangers, sharks, even Bigfoot! Their adventures made global superstars of Lindsay Wagner and Lee Majors, and the popularity of the bionic couple called the actors back several times over the decades to reprise their roles in made-for-TV reunion films.

It is in the stories of *The Bionic Woman* that the parable of the enhanced, evolved human really take shape. My childhood love for *The Bionic Woman* turned into full-blown hero worship in the early 90s, when through reruns on the SciFi channel, I rediscovered Jaime Sommers as a feminist, transhumanist role model for a new age. Here was the living, breathing embodiment of the union of science and nature, a woman whose super-strength meant she was nobody's victim, and whose education in psychology allowed for a nuanced, empathic approach to the people around her, whether friend or foe. Jaime was given an evolved physiology to go with an evolved consciousness, and she used it well. A super-strong and independent woman, Jaime knew just how powerful she could be, but she wielded it with grace, wisdom, and restraint. A reluctant secret agent on leave from her normal job as a teacher, Jaime exercises as much superhuman compassion in her missions as superhuman strength. As strong and as fast as I'd like to be, I'd like just as much to be as perceptive and kind as she is. Steve and Jaime teach us that the possibilities for human evolution lie not only in bionic enhancements, but in the knowledge of just how, and when, to use them.

ROBOCOP

ROBOCOP FRANCHISE, 1987

Subtlety, thy name is Verhoeven. Loud, brash, vulgar, and so very violent, *RoboCop* is a biting satire of American excess. It's as relevant now as it was in 1987 and quite prophetic considering the current political and economic climate of Detroit, where the film is set. Updating the Iron Man myth with a tragic human twist, RoboCop is a reluctant cyborg hero, a Terminator with post-traumatic stress disorder.

Crime-ridden Detroit of the near future is in a financial crisis, and the only solution is privatizing social services. Evil corporation Omni Consumer Products (OCP) has recently taken control of the police force, and is working on building a better, automated form of law enforcement on round-the-clock patrol. When their "enforcement droid" prototype ED-209 goes haywire and kills a board member during a demonstration, plans are set in motion for a machine with a little more reasoning power. A human/machine hybrid with the ability not only to wield the sword of justice, but balance the scales as well. All they need is a dead cop—enter family man and good guy Alex Murphy, who has the worst first day on the job ever. Killed by a ring of ultra-baddies in the pocket of OCP, Murphy is brought back to life as enforcement droid version 2, RoboCop. Harnessing the processing power of the human mind, OCP wipes Murphy's memory and programs him with "Prime Directives" to serve and protect—not just the citizens of Detroit, but corporate interests as well. It would've gone so well if it weren't for those pesky human memories, awakening the traumatic experience of being gunned down, setting RoboCop on a journey for vengeance. His fight against criminals and his creators contains the message that no matter how hard capitalism tries, people can never be made into products.

MOLLY MILLIONS

*JOHNNY MNEMONIC, 1981 &
NEUROMANCER BY WILLIAM GIBSON, 1984*

No list of robots is complete without paying homage to the work of the late-20th-century's greatest master of science fiction and father of the cyberpunk movement. Canadian writer William Gibson is the person who coined the term cyberspace, gave us the concept of the Matrix, and pioneered the wildly popular Steampunk genre. His 1984 novel *Neuromancer* sets the tone for his highly influential work and gives us one of the coolest cyborg assassins ever.

Gibson's universe is a post-apocalyptic, paranoid complex of overlapping synapses and blurred distinctions. Human is machine, artificial intelligence can become omniscient and "cowboys" ride the frequencies and synapses in people's brains, feeling their sensations and watching the world through their eyes. In Gibson's world, not a single human is without technological enhancement, humans can be turned into robots and death itself can be cheated by saving the collected memories and personality traits of a person onto a ROM file, where one can be accessed as a "Flatline Construct"—truly a ghost in a machine.

It's this bionic world in which we meet Molly Millions, possibly the most dangerous female cyborg in all of science fiction. She is, in a word, BADASS. Molly is a Razorgirl, a bionic mercenary implanted with physical and neural enhancements for combat. Lithe and agile, with a dancer's body and ninja reflexes, Molly is the ultimate gun-for-hire. She's got retractable blades under her fingernails, and mirrored glasses surgically implanted in her eye sockets. When she cries tears empty from her re-routed ducts into her mouth. Not that she cries much—her emotions are as steely as her hidden finger blades.

Molly has never been realized on screen but there's no doubt that she inspired Trinity in *The Matrix*, Aeon Flux, and Motoko Kusanagi. In the film adaptation of *Johnny Mnemonic*, Molly was replaced by a character named Jane, who had some of Molly's neural and assassin enhancements, but none of her Chrissie Hynde-inspired serpentine coolness. I'd love to see a proper film adaptation of *Neuromancer*, or have Molly show up in an all-new, Gibson-inspired franchise. It's about time Molly got hers.

MOTOKO KUSANAGI

GHOST IN THE SHELL, 1995

As much as you want to eff this lady, do not—I repeat—DO NOT eff with this lady. The sleek and sexy covert operative Major Motoko Kusanagi is not only the deadliest assassin in future Japan, she's the vessel for a brand new form of life, completely computer born. The heroine of manga and anime classic *Ghost in the Shell*, Motoko is Ripley, Rick Deckard, and Molly Millions all wrapped up into one intelligently designed, scantily clad package.

Motoko is the leader of Public Security Section 9 of the futuristic version of the Japanese National Public Safety Commission, a real-life organization that oversees the police force. This fictional division investigates cyber crime in a world that closely resembles William Gibson's *Neuromancer*. Motoko's unit fights a variety of cyber-villains including rogue hackers, evil human/machine hybrids bent on destruction,

and people whose memories have been implanted or corrupted. Differing in appearance slightly depending on which form of media you choose to indulge in, Motoko is a cool and lithe lady with a completely cybernetic body capable of astounding gymnastic feats; she sails through the air utilizing her "thermoptic camouflage" capabilities, completely invisible to the naked eye—and in the movie, completely naked. Like the other members of her team (and most people in this cyberpunk world), she's got neural inputs in the nape of her neck that allow her to tap into the web and telepathically communicate with mind and machine alike. If you're a fan of seeing the inner workings of robots (as I am), definitely check out the manga, as Motoko (and other cyborg characters) are frequently torn to pieces, requiring complete overhaul. My personal favorite is

the Mamoru Oshii film, a stunning and hypnotic anime classic, with hands-down the sexiest cyborg opening sequence ever detailing Motoko's construction. The film follows her fight against the "ghost hacker" The Puppet Master, and her existential quest as she wonders if she's even human at all or just a machine full of implanted memories. It's a beautiful, affecting film that has influenced a generation and is definitely my favorite of the anime genre. It also contains the most eloquent defence for the preservation of artificially intelligent life. When sentient cybernetic life form Project 2501 invades a body and demands political asylum with no threat of deactivation,

he makes the argument "that DNA is merely a program designed to preserve itself. Life has become more complex in the overwhelming sea of information. And life, when organized into species, relies upon genes to be its memory system. So man is an individual only because of his intangible memory, and memory cannot be defined, but it defines mankind. The advent of computers, and the subsequent accumulation of incalculable data, has given rise to a new system of memory and thought parallel to your own. Humanity has underestimated the consequences of computerization." Now how's THAT for a Robot Takeover?

DATA BYTE

Ghost in the Shell *was the first anime film ever to be released simultaneously in the UK, the USA, and its native Japan.*

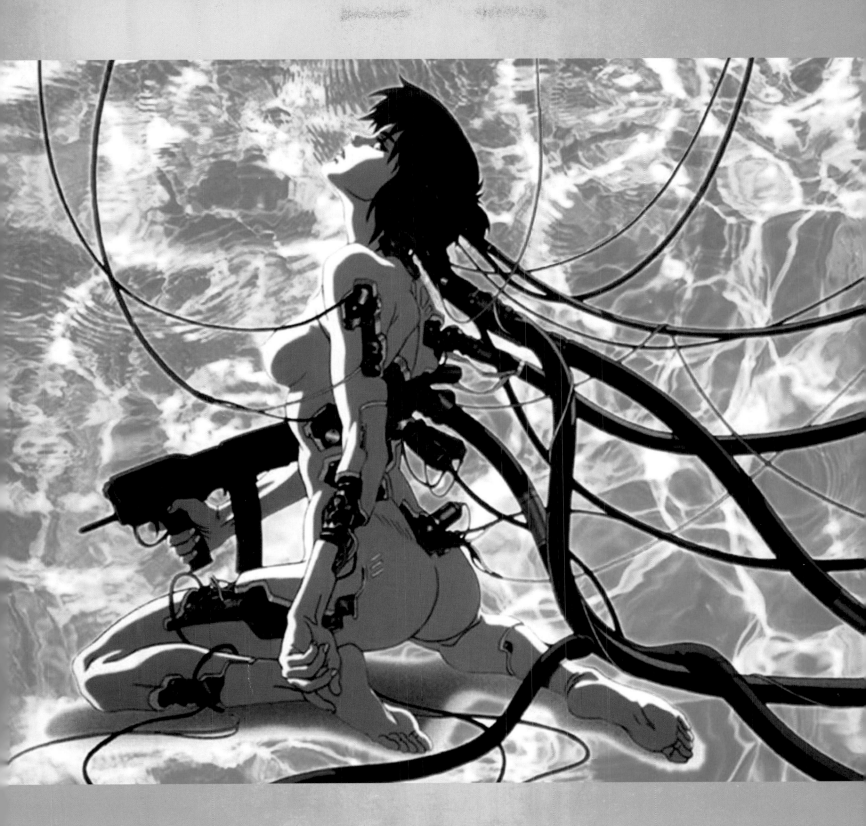

SEVEN OF NINE

A THREE-SEASON CHESS MATCH BETWEEN HUMAN AND BORG WILLS

Our next cyborg heroine began her story as a villain, and explores an interesting notion: can evil programming be undone? *Star Trek: Voyager* delves into this and many other interesting notions with Seven of Nine from the Borg Collective.

Assimilated at the age of six, young Annika Hansen was stuck in a Borg "maturation chamber" until she was old enough to serve as Seven of Nine, Tertiary Adjunct of Unimatrix Zero-One. She acts as the appointed Borg ambassador that Janeway and crewmate Tuvok must work with to defeat a common enemy and gain safe passage for the starship *Voyager*. After an uneasy alliance and an assimilation attempt, Seven of Nine is stranded on Voyager, her connection to the Borg Collective severed.

The question of what to do with their new shipmate brings up interesting questions of consent. Now that her connection is severed, should attempts be made to remove the Borg hardware, something that's been a part of her since age six? Seven of Nine insists she be returned to the nearest Borg vessel, but her human physiology begins to take over and reject the Borg technology, putting her life in danger. Janeway will not risk her crew to return the Borg drone, and believes the human underneath all the Borg hardware can be reached. This begins a three-season chess match between human and Borg wills as Janeway is determined to make Seven of Nine a valuable member of the *Voyager* crew. "We have something the Borg could never offer," she tells her First Mate, "friendship."

Seven's course of repatriation to humankind is a long, difficult education. She struggles with independence and isolation from the Collective. She has difficulty with authority and often gets in trouble with superiors as she circumvents them based on what she thinks is right. Though whenever called to put her Collective experience to use, she proves herself an excellent manager. She also encourages Janeway to forgo her own Starfleet "programming" in moments when out-of-the-box strategy is required. I consider Seven and Janeway's one of the strongest onscreen female relationships of all time, and watching tactical master Janeway untangle the knot of Seven's Borg conditioning is one of the more compelling storylines in robo-dom.

DATA BYTE

Seven of Nine's sexy skintight catsuit was in fact so restrictive that Jeri Ryan, the actress who played her, actually passed out a couple of times on set.

INSPECTOR GADGET

ANIMATED SERIES, 1983 & FILM, 1999

A CYBORG SWISS ARMY KNIFE OF VARIOUS HIDDEN APPLIANCES

A silly cyborg blend of detectives from such classics as *The Pink Panther* and *Get Smart*, Inspector Gadget was the co-production of American, French-Canadian, Japanese, and Taiwanese animation studios launched in 1983. The titular detective is a cyborg Swiss Army knife of various hidden appliances designed to make him the ultimate super sleuth. It's a shame they didn't extend the same upgrades to his intellect, as he's a pretty inept detective, but luckily his niece Penny and hyper-intelligent dog Brain are there to do the thinking for him.

Fighting Dr. Claw and a revolving coterie of his MAD crime organization cohorts, Inspector Gadget was a global hit, instantly iconic for his myriad malfunctioning attachments, his "go-go gadget" catchphrases delivered by Maxwell Smart (Don Adams) himself, and of course, the theme song, one of the most recognizable in the world. The success of the show resulted in three separate spin-off series, comic books, a Christmas special, and two full-length animated movies. Gadget got the Disney treatment in 1999 with a live-action feature film (pretty terrible) and a straight-to-video sequel (even worse), has had nine video games released since 1987, and was hilariously lampooned on *Robot Chicken*, given upgrades by the Cyberdyne (aka Terminator) Corporation, with Dr. Claw at the helm of Skynet. Inspector Gadget has recently been given a makeover in an all-new computer-animated reboot, ensuring everybody's favorite bungling bionic gumshoe will go-go gadget well into the 21st century.

"GO-GO GADGET!"

– INSPECTOR GADGET

DATA BYTE

Although the Inspector Gadget theme tune has now become iconic, copies of the original television soundtrack are now extremely rare. If you have one, keep hold of it!

MARCUS WRIGHT

COMBINING THE DEADLY HARDWARE OF THE TERMINATORS WITH THE PERFECT INFILTRATION COVER OF THE HUMAN BODY

The *Terminator* franchise has given us some very memorable robots over the years, mostly in the guise of the time-traveling humanoid assassins. *Terminator Salvation* put us right in the middle of the War Against the Machines, finally immersing audiences in the world after Judgement Day. With this new world came a new kind of machine created by the fascist Skynet, one it was sure could wipe out the Resistance once and for all. One who was completely unaware he was a weapon at all. He's the Unknown Cyborg, Marcus Wright.

We first meet convicted cop-killer Marcus in his cell on Death Row, donating his body to the Cyberdyne Corporation. The next time we see him, it's 15 years later, the War Against the Machines is in full swing, and John Connor passes over his inanimate body on a slab in a Skynet outpost. After the hidden base is blown up, Wright wakes up (is activated?) and emerges, a muddy and hollering golem totally new to this blasted world. Through the course of the film he discovers what has happened to the planet, and what Skynet has done to him. In a plot that echoes the Philip K. Dick stories "Second Variety" (page 42) and "Imposter," Marcus discovers that he has been made a tool for the Machines. Part of a new research and development effort, Marcus is the prototype for the T-H, or hybrid, combining the deadly hardware of the Terminators with the perfect infiltration cover of the human body. He proves to be the perfect covert agent, but not the one Skynet had in mind. This bulletproof warrior puts his cybernetics to work to double-cross the evil machines, and proves that Skynet, even with all its power, will never hack humanity: "you can't put it into a chip. It's the strength of the human heart, the difference between us and machines."

DATA BYTE

Christian Bale was originally supposed to play Marcus Wright but ended up playing the character of John Connor.

ROBOTS

REALIZED

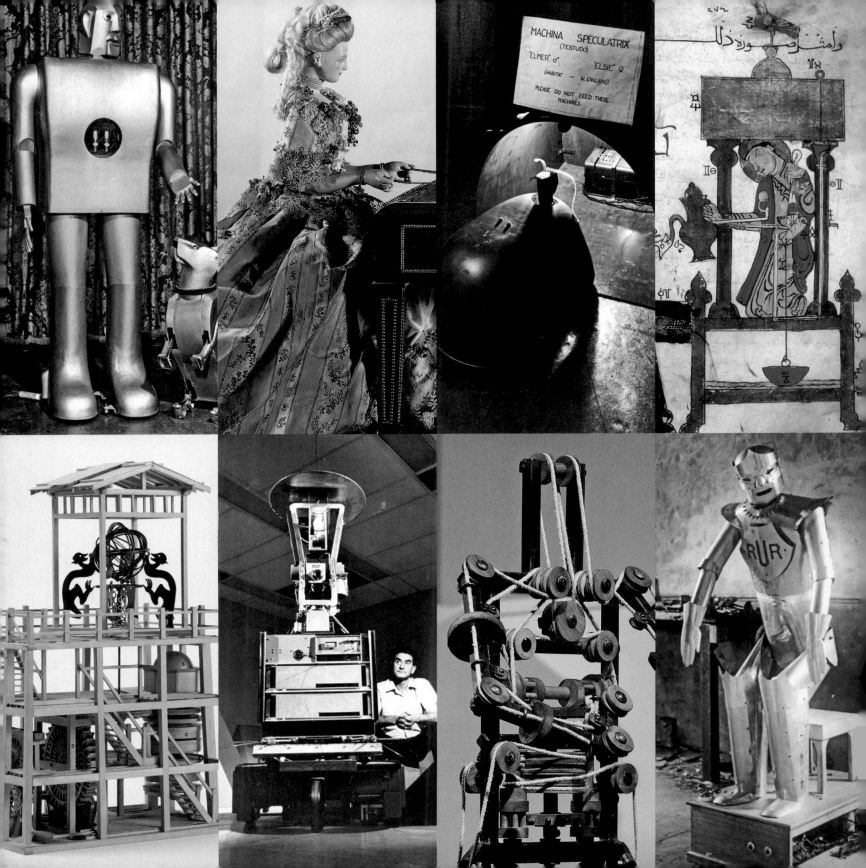

EARLY PROTOTYPES

We've reached the end of one chapter and the beginning of another—one that may not be as fantastical and far out as the last, but is certainly just as exciting. As I attempted to illustrate in the first part of this book, the idea of robots as animated toys or tools has been bouncing around in the human mind since the dawn of thought. As we look at how robots have been realized in our everyday lives, so we must also go back to the very first attempts at automating computation and emulating the mechanics of the human body.

Automatons exist in historical record as far back as ancient Egypt, have been written about in the Mahabarata and the Bible, and were designed and described by Hero of Alexandria, the inventor, teacher, mathematician, and author often considered the world's first cyberneticist. In this first section of Robots Realized we will move through nearly a thousand years of history, examining the early prototypes from the first "computer," to the first creations ever publicly referred to as robots, to the first artificially intelligent superstars of the 20th century. It's time to get real.

SU SONG'S COSMIC ENGINE

ELEVENTH CENTURY

THE FIRST-KNOWN USE OF A CHAIN DRIVE, WHICH SU CALLED "THE CELESTIAL LADDER"

As technology advances, it miniaturizes. Computers that used to fill an entire room can now fit easily into our pockets. As we look back on history, our computers get more and more primitive, and a whole lot bigger.

How big is Su Song's Cosmic Engine? Five stories. A six-year project headed by the renowned engineer, scientist and diplomat of China's Song Dynasty, this hydro-mechanical wonder was completed in Kaifeng around 1092 AD, predating Europe's first astronomical clock at Strasbourg Cathedral by well over two centuries. It featured the world's first automated armillary sphere and celestial globe, some of the first uses of escapement technology (the gears that make a clock tick) and the first-known use of a chain drive, which Su called "the celestial ladder." The clock also used a five-tiered rotating cast of mechanical mannequins, armed with percussion instruments or plaques, which would move into view along the tower's façade to announce the time and sound the hours.

This proto-computer not only told the time, but also measured the weeks, months, seasons, the movement of the stars, and the position of the sun. An invaluable tool for Emperor Zhezong's astronomers and astrologers, the Cosmic Engine was sadly dismantled by the invading Jurchen army of Manchuria in 1127. Though detailed diagrams existed, the clock could not be reassembled, and the over 400 parts have been lost to history.

There are quite a few scale models on display around the world including one at London's Science Museum; two life-sized replicas exist at the Gishoda Suwako Watch & Clock Museum in Japan's Nagano Prefecture and one at the National Museum for Natural Science in Taiwan. A marvel of ancient technology and one of the world's first automated multi-function machines, think of Su Song the next time you check your watch.

DATA BYTE

Su Song was a talented man: as well as a scientist he is credited as being a mathematician, statesman, architect, and poet, among other skills.

AL-JAZARI'S INGENIOUS DEVICES

TWELFTH & THIRTEENTH CENTURIES

A HIGHLY TECHNICAL AND SKILLED CRAFTSMAN WHO APPLIED HIS KNOWLEDGE TO THE DESIGN OF A VERITABLE GARDEN OF MECHANICAL DELIGHTS

While Europe suffered years of economic and scholarly regression in the Dark Ages, the Muslim world was going through a cultural and intellectual enlightenment that would transform history. A host of diverse innovations were made that modern civilization continues to employ including windmills, alcohol distillation, shampoo, coffee, supermarkets, and the three-course meal. It was the age in which the scientific disciplines of algebra and chemistry were born, and it gave us the first humanoid robots—not just automatons made to entertain, but ingenious inventions intended to mechanize and improve daily life.

Badi' al-Zaman Abu al-'Izz ibn Ismail ibn al-Razzaz al-Jazari, or simply Al-Jazari, is the proto-Renaissance Man of the late 12th century. Named for his birthplace in southeast Turkey, Al-Jazari worked, like his father before him, for the Artuqid kings of Diyarbakir on various engineering projects. A highly technical and skilled craftsman, Al-Jazari applied his knowledge to the design of a veritable garden of mechanical delights. Pictured left is his design for a hand-washing basin that utilizes the flush mechanism used in modern toilets. Pulling a lever causes the water to drain from the basin, while a female automaton refills it from her pitcher. Another more sophisticated version featured a peacock that would fill the basin from its beak and trigger a hidden servant to pop out with soap and towels. There were automata that served drinks, elaborate reprogrammable

astronomical clocks that used water or burning candles to power their mannequins, and a boat full of mechanical musicians that would play different rhythms by switching out their cams—a 13th-century drum machine! Al-Jazari compiled these designs into the *Book of Knowledge of Ingenious Mechanical Devices* in 1206, leaving detailed instructions on how to create over a hundred amazing inventions for future generations. A touring exhibition created by the 1001 Inventions organization features his 16-foot (5-meter) high, water-powered timekeeping marvel the Elephant Clock, as well as a host of other mechanical innovations of the Islamic world. The exhibition is constantly on the move; check their website at 1001inventions.com to see if you might have a chance to get up close and personal with the genius of Al-Jazari.

DATA BYTE

Al-Jazari's 800-year-old Elephant Clock had features of Indian, Chinese, Greek, Arabian, Egyptian, and Persian origin, in a deliberate celebration of diversity.

LEONARDO'S MECHANICAL KNIGHT

FIFTEENTH CENTURY

LEONARDO DA VINCI WAS WHOLLY INVESTED IN CREATING MACHINES TO IMPROVE OUR LIVES

The original Renaissance Man designed an entire museum's worth of ingenious contraptions throughout his life. From war machines to hydraulic saws to moving theatrical scenery and programmable mobile carts, Leonardo da Vinci was wholly invested in creating machines to improve our lives.

One of history's keenest observers of the natural world, Leonardo also created machines that mimicked the designs found in nature, including his flying machines based on birds and bats, and a mechanical lion for King Louis XII of France. Leonardo applied his studies of anatomy and kinesiology to his project known as the Mechanical Knight, designed in 1495, just a few months before he would start work on *The Last Supper*.

Operated by an external crank, the Mechanical Knight was a suit of armor with a system of gears and pulleys that would control its movements. Two separate mechanisms powered the upper and lower halves of the body with joints at the ankles, knees, and hips down below and the hands, wrists, elbows, shoulders, and jaw up top. It could sit, move its head from side to side, wave, and lift its visor.

Plans for the Mechanical Knight were lost until the 1950s, when scholar Carlo Pedretti pieced together several fragments of Leonardo's notes. Roboticist Mark Rosheim studied these fragments in the 1990s and made a working model of the Mechanical Knight, now fully motorized. Rosheim applied Leonardo's principles to his own "anthrobots," or humanoid robots, working for NASA today, like Robonaut whom we'll meet in just a few page turns (page 200). For more on Leonardo's fantastic inventions and proto-robots, you can read Mark Rosheim's book *Leonardo's Lost Robots*.

DATA BYTE

Leonardo's mechanical knight was built for a pageant through the streets of Milan, possibly the wedding of the Duke of Milan's niece.

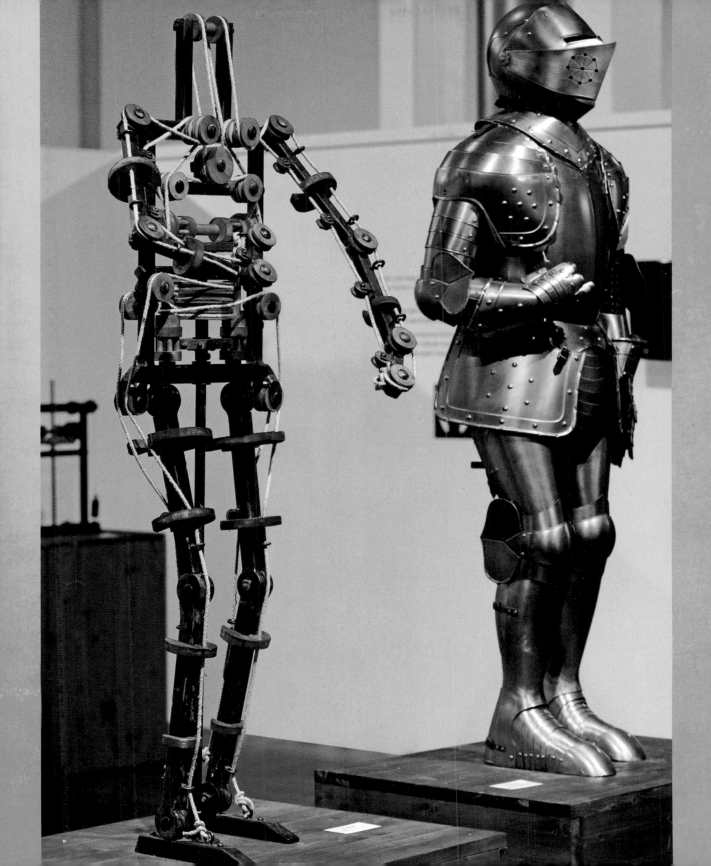

THE AUTOMATON CRAZE

EIGHTEENTH CENTURY

You'll have to forgive me: in the interest of cramming in as many cool robots as I could, I've taken the liberty of lumping together a century's worth of amazing automatons. Popular since the dawn of civilization, automatons are the embodiment of the joy and wonder of our natural world, an expression of skilled craftsmanship, and represent some of the world's first music on demand. Popularity peaked in the 18th and early 19th centuries, with clockwork creations traveling the globe, inspiring great minds and even causing a scandal or two.

In ancient Greece, the word automaton described any self-propelled machine and evolved to mean a machine in imitation of life as man's world view became more mechanistic. The eighteenth-century craze for automatons was sparked by French mechanic Jacques de Vaucanson and his three incredible creations: a life-sized flute player that could play 14 different tunes; a fife-and-drum player; and a duck. Vaucanson's Duck could walk, swim, preen its feathers, and even eat, drink, and defecate thanks to its separate internal compartments. It was a massive hit in Paris and wondrous automatons began appearing all over Europe.

Swiss father-son duo Pierre and Henri-Louis Jacquet-Droz demonstrated their skills in the 1770s with The Musician, The Draughtsman, and The Writer, the latter of which had a device to control what the automaton writes. James Cox's Silver Swan can still be seen at the Bowes Museum in England and the Peacock Clock at the Hermitage Museum in St. Petersburg, Russia.

Other famous automata include Henri Maillardet's poetry and picture-drawing writer (the inspiration behind *The Invention of Hugo Cabret*), and the chess-playing Turk invented by Wolfgang von Kempelen. This one actually caused quite a stir: the turbanned figure of a man seated at a chess board, audience members could play a game with the Turk and many were famously beaten, including Napoleon Bonaparte. What was not apparent to audiences was the concealed chess master hiding inside the cabinet, moving the Turk's arm to make the next play—or in Napoleon's case, knocking all the pieces off the board when the French ruler tested it with illegal moves. This false automaton inspired ETA Hoffman's iconic stories "The Sandman" and "The Automaton" as well as the 1893 horror classic "Moxon's Master" by Ambrose Bierce. In turn Edgar Allan Poe wrote not one but two treatises on its workings. An 1820 match with the Turk influenced Charles Babbage's dream of making his own true game-playing machines, resulting in the world's first automatic computers, the Difference and Analytical Engines.

DATA BYTE

Peter Kintzing and David Roentgen collaborated to create the dulcimer-playing automaton said to be based on Marie-Antoinette, who bought it in 1784.

STEAM MAN

NINETEENTH CENTURY

THE WORLD'S FIRST SELF-PROPELLED VEHICLE

Though the image to the right is a fictional depiction, *The Steam Man of the Prairies* was based on a very real mechanical man from New Jersey.

Created by Zadoc Dederick and Isaac Grass of Newark, the Steam Man was the world's first self-propelled vehicle, a robotic rickshaw driver intended to usher in the era of the horseless carriage. Standing at an imposing 7 feet 9 inches (2.36 meters) tall, this mustachioed gentleman carried a three horsepower steam engine in his chest cavity whose fumes would exit through the chimney concealed in his top hat. Additional machinery on the Steam Man's back was hidden under a knapsack and blanket roll, giving him a refined yet rough-and-ready appearance.

Capable of speeds of up to a mile a minute (1.6 kilometers per minute), the Steam Man was unfortunately terrible on an incline, and according to the *Newark Observer*'s article of 23 January 1868, any obstacle "more than nine inches below or above the level of the road" could cause problems for the Steam Man.

A patent was awarded to Dederick in 1868, but sadly he never realized his dream of the streets running rampant with Steam Men, though his invention did spark quite a trend. The wonder of walking machines carried on with other Steam Men including the Thomas J Winans/Joseph Eno "Steam King," CC Roe's "Adam Ironsides" in Canada and Professor George Moore's Steam Man, made to resemble a medieval knight.

The literary *Steam Man of the Prairies* by Edward Ellis was published in 1868. It was the first science fiction dime novel published in the United States and would go on to inspire L Frank Baum to create Tik Tok, the Iron Man of the Land of Oz (page 77) and Paul Guinan's Steampunk hero Boilerplate (page 92).

DATA BYTE

Dederick gave the Steam Man a human-like appearance and dressed it in clothes in order to prevent it from frightening horses out on roads.

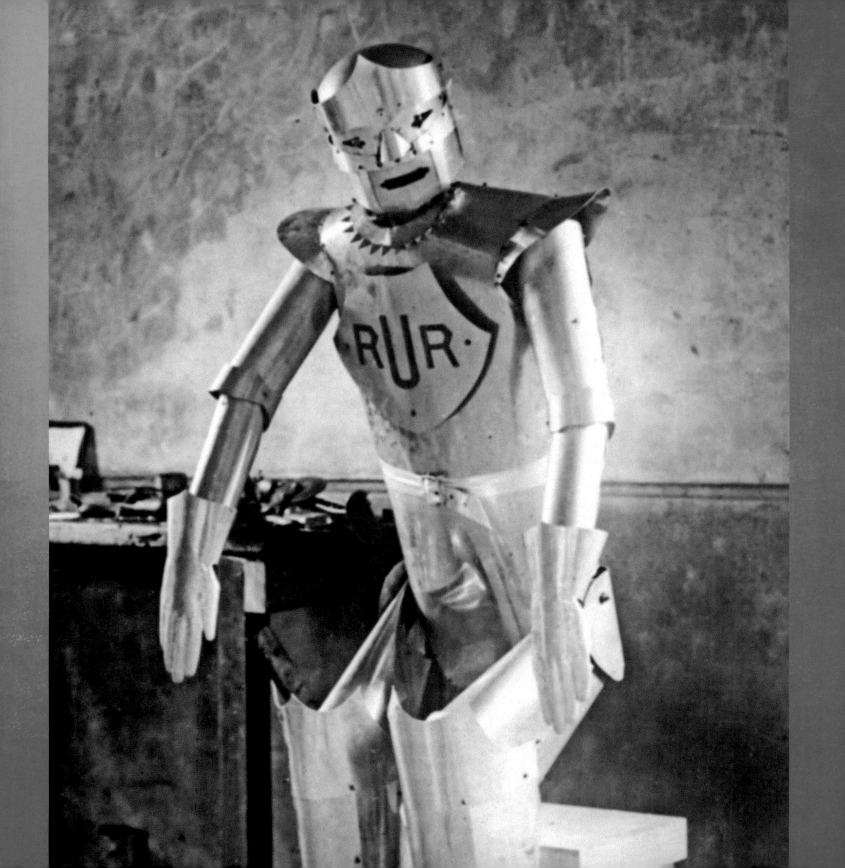

ERIC

ENGLAND'S QUINTESSENTIAL MECHANICAL GENTLEMAN

Karel Capek's seminal play *RUR* caused a sensation all over Europe and gave the world the word that perfectly defines the modern mechanical man. In 1923 *RUR* was published in the UK, where the action in the play takes place, and the word robot quickly entered the English language. Just five years later, England would debut its very first robot prototype in the form of Eric.

Before the creation of Eric WWI veteran and journalist Captain William H. Richards, who was also secretary of the Society of Model Engineers, was in a quandary. The Society needed somebody to deliver the commencement speech for its upcoming exhibition in London as the originally scheduled guest, the Duke of York, couldn't attend. Richards needed a knight in shining armor to save him and dazzle the audience—so he decided to make one. A six-foot (1.82-meter) tall automaton in a suit of polished aluminium armor, Eric was even outfitted with a little RUR branding on his chest in homage to Capek's play. Constructed by Richards and engineer A.H. Reffell, Eric's feet were attached to a box that concealed a 12-volt electric motor. Another was inside his chest along with 11 electromagnets and nearly three miles of wire to control the movements of the upper torso. His eyes were bare white bulbs painted with red pupils, and blue sparks spit out of his mouth when his jaw moved. Eric was unveiled on September 20, 1928, at the Royal Horticultural Hall, where he rose from his seat, bowed, looked to the right and left, waved, and delivered a four-minute welcoming speech to the assembly. Eric was able to answer questions from the crowd, tell the correct time when asked, and would stop speaking when given the command to "shut up." An offstage orator delivered Eric's lines through a hidden speaker in his chest, but Captain Richards refused to tell audiences how it was done, cryptically revealing that it was "neither phonograph or talking film." Eric went down a storm and W.H. Richards travelled the world showing him to audiences in Europe, America, and Australia. Eric is the world's first live-and-in-person creation to be referred to as a robot, cementing this shining electric image in minds the world over and inspiring the likes of Fritz Lang and the Binder brothers to put pen to paper. Captain Richards' creation will live on as England's quintessential mechanical gentleman and our first glimmer of the dawn of the Age of the Robots.

DATA BYTE

Eric proved so popular that Captain Richards even created a follow-up, new-and-improved model named George whom he continued to travel the world with.

GAKUTENSOKU

"IF ONE CONSIDERS HUMANS AS THE CHILDREN OF NATURE, ARTIFICIAL HUMANS CREATED BY THE HAND OF MAN ARE THUS NATURE'S GRANDCHILDREN" — MAKOTO NISHIMURA, *SUNDAY MAINICHI MAGAZINE*

Japan has enjoyed a long love affair with the robot, without the fear and skepticism inherent in Western culture. This may stem from the centuries' old tradition of depicting people in mechanical form, from Bunraku Puppet Theatre to the tea-serving Karikuri automatons in the 18th century, or from the dominant Shinto religion that ascribes a divine essence to Earth's natural creations. Objects like trees or rocks are very much alive, possessing a unique *kami* or spirit. Couched in these terms, it's not a stretch to fathom that those who conceive of a "rock-spirit" can embrace the concept of a "robot-spirit" without fear (a notion expressed by Daniel H. Wilson). Have a look back at the section Murderous Malfunctions and Fascist Machines (pages 36–55) and you'll notice that not one of those robots comes from Japan. It's in this spirit that Japan's first real *robotto* was introduced in 1929, a creation whose name means "learning from the laws of nature."

Gakutensoku, like Eric before him (page 161) made reference to the robots-in-revolt of *RUR* (page 38), but in a different way. According to novelist Hiroshi Aramata, Gakutensoku was "an attempt to set aesthetic robots free from slaves to industry." First exhibited in Kyoto as part of the celebration of the new Emperor's rise to the throne, Gakutensoku was a 10-foot (3-meter) tall automaton created by Makoto Nishimura in Osaka. Gakutensoku used an internal air pressure system to change the expression on his gold rubber face, puff out his cheeks and chest, and move his head and arms. In his left hand Gakutensoku held a lamp called *Reikanto* (meaning "inspiration light") and in his right hand grasped a pen. When the lamp would switch on, the automaton wrote words in Chinese calligraphy; when the robotic bird above his head cried, Gakutensoku closed his eyes in what spectators would call a "pensive" or "nostalgic" expression. Gakutensoku was displayed all over the world but was sadly lost during a tour of Germany in the 1930s. The Osaka Science museum built a full-size and fully functioning replica of its robotic native son in 2008, where he is on permanent display huffing, puffing, and writing in devotion to the spiritual essence of nature.

DATA BYTE

An asteroid—9786 Gakutensoko—was named after the robot after it was discovered in 1995.

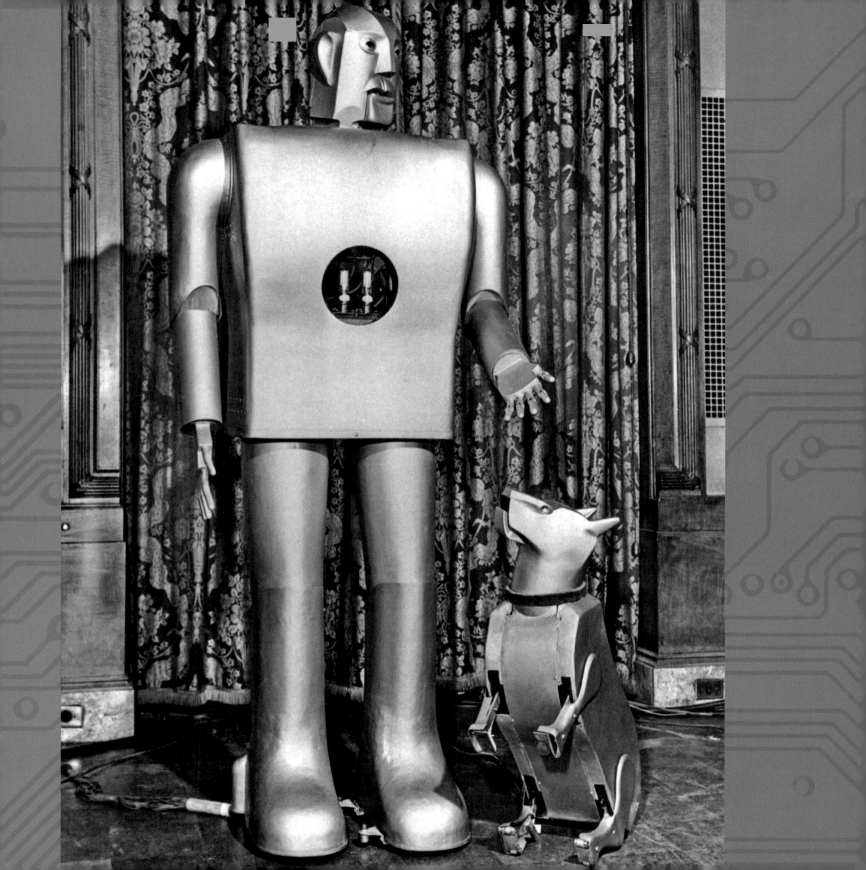

ELEKTRO & SPARKO

1930S

THE WALKING, TALKING ROBOT STAR OF THE 1939—40 WORLD'S FAIR

The 1920s and 30s brought an explosion of experimentation in humanoid robot prototypes. Engineers in Europe, Japan, and the US tested the latest light and radio technology in mechanical men, many of which became local celebrities. Westinghouse Electric Company's output began in 1926 with Roy J. Wensley's Herbert Televox and culminated spectacularly in the walking, talking robot star of the 1939–40 World's Fair, Elektro the Moto-Man and his dog Sparko.

The theme of the fair was "Dawn of a New Day" and promised visitors a glimpse of the "world of tomorrow." Exhibits introduced spectators to many wondrous new technologies including television, air conditioning, color photography, and fluorescent lighting. Elektro the Moto-Man was designed by Westinghouse's resident engineer Joseph M. Barnett. 7 feet (2 meters) tall and 248 lbs (112 kilograms), Elektro contained 11 motors, 48 relays, and the same kind of photocells and electric eyes used for automatic doors. These light and color sensors meant he could tell the difference between red and green, and a microphone allowed his operator to deliver simple voice commands to control his 26 different movements. This included walking, turning his head, moving his jaw, and counting on his fingers. A hidden bellows allowed the Moto-man to smoke cigarettes and also gave him the kiddie-pleasing ability to inflate balloons. He could speak about 700 words via an internal 78 rpm turntable and converse with audiences the same way Eric (page 161) would, with an offstage "man behind the curtain" answering questions and delivering one-liners. The following year brought Sparko, designed by Don Lee Hadley in homage to his Scottish terrier, who would sit, beg, turn his head, and bark at the command of his master's voice. Over 44 million people attended the 1939–40 World's Fair and Elektro and Sparko continued entertaining crowds through the 1950s. Elektro appearanced as the campus computer "Thinko" in Mamie Van Doren's *Sex Kittens Go to College* (1960), and was sampled in the 1992 song "Original Control (version 2)" by Meat Beat Manifesto. Though Elektro was sent to the scrapheap and his head given to a retiring engineer in the 60s, the Moto-Man was restored in 2006 and he and Sparko can be seen today in permanent exhibition at the Mansfield Memorial Museum.

DATA BYTE

At the World's Fair, Elektro made statesman-like appearances "speaking" to the onlookers below from a balcony overlooking the fair.

DISNEY'S HALL OF PRESIDENTS

1971

NOBODY CONTRIBUTED TO THE ANIMATRONICS PHENOMENON MORE THAN THE GREAT WALT DISNEY

Like the automaton, animatronics are proto-robots, lifelike puppets with a certain set of preprogrammed movements, but unlike their clockwork cousins these models have the added benefit of electricity.

A portmanteau of *animated electronics*, the term was coined in the 1960s to define the very realistic electronic mannequins used in film special effects and live entertainment, and nobody contributed to this phenomenon more than the great Walt Disney. Animatronics helped Disney bring to life the different characters and worlds of his great theme park opened in 1955, and in 1964 he was the first to use animatronics in film for *Mary Poppins*. That same year Disney debuted the world's very first talking, moving copy of a human being in an attraction that would go on to inspire equal parts wonder and fear in generations of visitors.

Original plans for the Hall of Presidents called for the US leaders to be immortalized as wax figures, but Disney the dreamer knew he could give audiences more and set to work with his "Imagineers" to bring history to life. He began by building a prototype of his favorite president and boyhood hero, Abraham Lincoln. When famed New York city planner Robert Moses visited Disney in 1962 to get ideas for the upcoming World's Fair, he saw the prototype and enlisted Disney to create the *Great Moments with Mr. Lincoln* stage show for the Illinois Pavilion. The World's Fair opened in Flushing, Queens, on 22nd April 1964, and the first ever audio-animatronic star was born. Audiences watched the life-sized Lincoln talk, gesture, and rise from a chair delivering a "medley" of his greatest speeches. A second version of the stage show opened at Disneyland in 1965 and expanded to Disney World in 1971, featuring every single US President as well as an animatronic Benjamin Franklin, the first ever with the ability to climb stairs.

In 2009 Disney debuted a brand new autonomatronic President Lincoln using the next generation of animatronics that replaces the traditional hydraulics with computerized controls, allowing the figure to "choose" its movements and interact with the audience. The first ever robot with a human face, Disney's innovation is preparing us for the very real possibility of very real-looking humanoid robots in the future.

167

EARLY PROTOTYPES

DATA BYTE

After Barack Obama's inauguration in 2009 the Hall of Presidents was updated with Morgan Freeman providing the narrator's voice.

ELMER & ELSIE

1948

We switch gears now from the automatons that only thought about autonomy to the first pair of self-propelled, decision-making robots. In the late Forties at the Burden Neurological Institute in Bristol, England, William Grey Walter was studying the human brain. He believed only a small number of cells were required to bring about a host of complex behaviors, which he proved in an experiment with two simple machines: Elmer and Elsie, two "Machina Speculatrix" that displayed the first-ever independent movement in a robot and may also have displayed the first instances of robotic self-awareness.

Constructed between 1948 and 1949, Elmer and Elsie were "free goal-seeking mechanisms" outfitted with light-sensitive photoelectric cells that would rotate above their hard plastic shells. Light stimulus would activate the motor, moving each three-wheeled "tortoise" forward toward its luminous goal. Sensors allowed Elmer and Elsie to maneuver around any obstacles and retain a short memory of their placement and when power ran low, they could find their own way back to their recharging "hutch." If two light goals were set up, the tortoise would "decide" which goal to move toward, prompting Walter's conclusion that "by the scholastic definition the creature appears to be endowed with 'free will'. It approaches and investigates first one goal and then abandons this to investigate the other." In another experiment, Walter set up a mirror and attached a small flashlight to the front of Elsie. As the tortoise moved toward the reflected light in the mirror, it "saw" its reflection, which caused it to start "flickering, twittering, and jigging," prompting Walter to call it his "clumsy Narcissus" and write that if the same behavior were observed in an animal it "might be accepted as evidence of some degree of self-awareness."

The fate of the original Elmer and Elsie is unknown, though it is thought that technician W.J. "Bunny" Warren used them as parts when creating six new tortoises for Grey Walter in 1951. This next generation was displayed at the Festival of Britain and traveled the world throughout the 1950s. The same technology pioneered by Elmer and Elsie led to the development of BEAM robotics and is still widely used in many self-propelled machines, including the most popular domestic robot working on the planet today, who we'll meet soon.

SENSORS ALLOWED ELMER AND ELSIE TO MANEUVER AROUND ANY OBSTACLES

DATA BYTE

Some of the follow-up tortoises to Elmer and Elsie are still around today. One is owned by the Smithsonian Institution and another one is on permanent exhibition at the Science Museum in London.

ROBOT CONVERGENCE

172 I have been trying to whet your whistle with all things robotic, so what to do if you've developed an unquenchable thirst? What if it's not enough to watch, read, and think about robots, and you've gotta get personal? Robot lovers have many options to converge at places around the world. From hack-a-thons and competitions to meet-ups for enthusiasts bent on building their own Robby or B-9 to large-scale conventions with corporate sponsors and expos galore, the world of the Roboconvergence is vast and just waiting for you to join in the fun.

The best-known robot convergence is the Robot Soccer World Cup or RoboCup, founded in 1997 and committed to promoting advancements in AI at a globetrotting annual affair described by roboticist Dennis Hong as a "research event packaged as a more exciting competition." Teams of roboticists from around the world gather to show off their AI software and hardware innovations and pit them against one another in soccer tournaments both virtual and real. There are different leagues in the RoboCup for different applications including Home Robots, Rescue Robots, and the soccer tournament has games for small and medium-sized robots and a Standard Platform League (see facing page) where teams use the same autonomous robot body (Aldebaran's Nao) and pit their software against one other to see whose algorithm is the most intelligent. There is even a Junior League for students with soccer, dance, and rescue competitions. The goal of the entire event is to produce a humanoid robot that will win a FIFA-compliant game against the current World Cup Champion team. They think this can be achieved by the middle of the 21st century—so get ready for the reality of a robo-Ronaldo bending it like Beckbot.

If getting to know the people who design, code, manufacture, and think about robots is something that interests you, there are two prestigious organizations you should know about, the International Symposium on Robotics (ISR) and the Robotics and Automation Society (RAS), part of the Institute of Electrical and Electronics Engineers (IEEE). Both organizations publish journals and stage conferences for their members internationally—you can check their respective websites to look for the sponsored event in your neck of the woods. Robots, their makers, and the people who love them come together at corporate-sponsored events covering the business, manufacturing, or personal application of robots. The RoboUniverse expo is the "first global tradeshow series for the service robotics industry" taking place in New York, San Diego, Seoul, and Tokyo where the public can interact with the newest innovations in robotics on the expo floor and also attend tutorials and seminars to learn about the latest in home, healthcare, and industrial robots.

If you'd rather see the greats of the robot world and don't mind a velvet rope or plate of glass in front of you, head to Pittsburgh to the Carnegie Science Center and visit roboworld™, the planet's largest permanent exhibition of robots that includes the Robot Hall of Fame display. There is also the museum at MIT, the world's first robotics lab in Cambridge, Massachusetts and the Tech Museum of Innovation nestled in the heart of Silicon Valley in San Jose, California where kids can build and interact with their own simple robot and many other AI exhibits. If you'd like to see one robophile's amazing

collection and replica work, head to Washington state and get in touch with John Rigg and make an appointment to see his incredible Robot Hut. There are science museums in every major city in the world and robotics exhibitions are always a crowd-pleaser—check with your nearest one to find out if any robots will be making an appearance in your area.

If all else fails, there's always Disneyland. Yes, their state-of-the-art animatronics are getting more and more lifelike and interactive, but they've also taken to showing off the present and future of robotics in a show where audiences can get up close and personal with the world's most well-known real robot, who will usher us in to the final section of Robots Realized and the remaining robots on our list of 100 Greatest.

THE STANDARD PLATFORM LEAGUE USED TO BE THE FOUR-LEGGED LEAGUE, UNTIL AIBO, THE ROBOT DOG, WAS REPLACED BY NAO

ASIMO

2000

174

If you're lucky enough to attend one of the aforementioned robot conventions, you also may have the good fortune to meet the world's best-known robot ambassador. Honda's Advanced Step in Innovative MObility or ASIMO pays homage to the great Isaac Asimov not just in name, but also living up to his legacy by being the most advanced bipedal robot on the planet.

Honda kept their walking robot mission top secret, from their very first E Series humanoid robot built in 1986 (essentially a box on legs), all the way to the first P Series model in 1993 that added an upper torso. P2 was unveiled to the public in 1996 as the first fully autonomous bipedal robot, and with its TV-set head, looked like something straight out of 1960s science fiction. The design was further streamlined into the classic friendly spaceman we all know and love today, the model P3, unveiled in 2000. A fully autonomous robotic assistant for the home and workplace, ASIMO is Honda's hallmark of innovation and the most instantly recognizable real robot today.

At just over 4 feet (1.2 m) tall, ASIMO is sized to assist a seated adult and perform tasks like switching on lights, pushing a cart, or opening doors. Wifi capabilities enable ASIMO to look up simple things online, tell you the weather, or coordinate appointments. ASIMO's hands have four fingers and a thumb just like a human and can perform highly specialized tasks like opening and pouring bottles and communicating in both American and Japanese Sign Language.

ASIMO has face and voice recognition, can distinguish between multiple people speaking at once, and recognize their gestures as well as the position, direction, and distance of multiple moving objects. It can work in conjunction with others of its kind to perform group tasks, and with its predicted movement control can avoid approaching humans or vehicles and return to its charging station when batteries run low.

ASIMO's mobile capabilities are truly what set it apart from the pack: its i-Walk and Zero Moment Point controls give the robot greater stability and enable ASIMO to prevent falls, sidestep obstacles, or maneuver up and down stairs—just ask the Daleks how difficult that is! The standard by which humanoid robots are measured, ASIMO continues to demonstrate what is possible with robotics right now—a present that looks a whole lot like the future Asimov envisioned.

ASIMO IS SIGNIFICANT AS THE FIRST ROBOT CAPABLE OF RUNNING, AND IS THE FASTEST HUMANOID ROBOT IN THE WORLD

DATA BYTE

The ultimate robot diplomat, ASIMO has toured the world and has greeted visiting luminaries to Japan, including Angela Merkel and Barack Obama, who got a special demonstration of ASIMO's soccer skills.

DA VINCI SURGICAL SYSTEM

1999

HIGHLY SPECIALIZED INSTRUMENTS, CAPABLE OF PERFORMING EXTREMELY PRECISE MOVEMENTS

Surgery, when medically necessary, can be almost as traumatic as disease itself. To make an incision large enough for a surgeon's hands to work within the body requires a great deal of healing time once the procedure is finished and can leave large unsightly scars.

Laparoscopic surgery, introduced in the 1980s, allowed surgeons to use finer, longer instruments that required much smaller incisions. The only problem was that these instruments were not very dexterous, and required the surgeon to look up and away from the patient at a monitor to see what was going on. Enter the field of telerobotics and the da Vinci Surgical System. Introduced in 1999, this surgeon-operated robot improves upon laparoscopic technology (and the human body) by using instruments that have a greater range of motion than the human wrist and an operating console with a high-definition camera that allows the surgeon to see everything in front of them in 3D. The highly specialized instruments allow for multiple small incisions instead of one large one and are capable of performing extremely precise movements, like sewing a bypassed artery onto a beating heart without cracking the ribs or opening the chest. This minimally invasive technique reduces risk, speeds healing time for the patient, and has the potential to take the surgeon to operating rooms around the world without having to get on a plane.

In 2001 the first transatlantic telesurgury took place, with doctors in New York successfully removing the gall bladder of a 68-year-old woman in Strasbourg, France, cementing the da Vinci Surgical System as the herald of a robot renaissance in the field of medicine.

DATA BYTE

The da Vinci Surgical System was incorrectly named after the great thinker who was only known as da Vinci because of where he was from. Academics always refer to him as simply Leonardo.

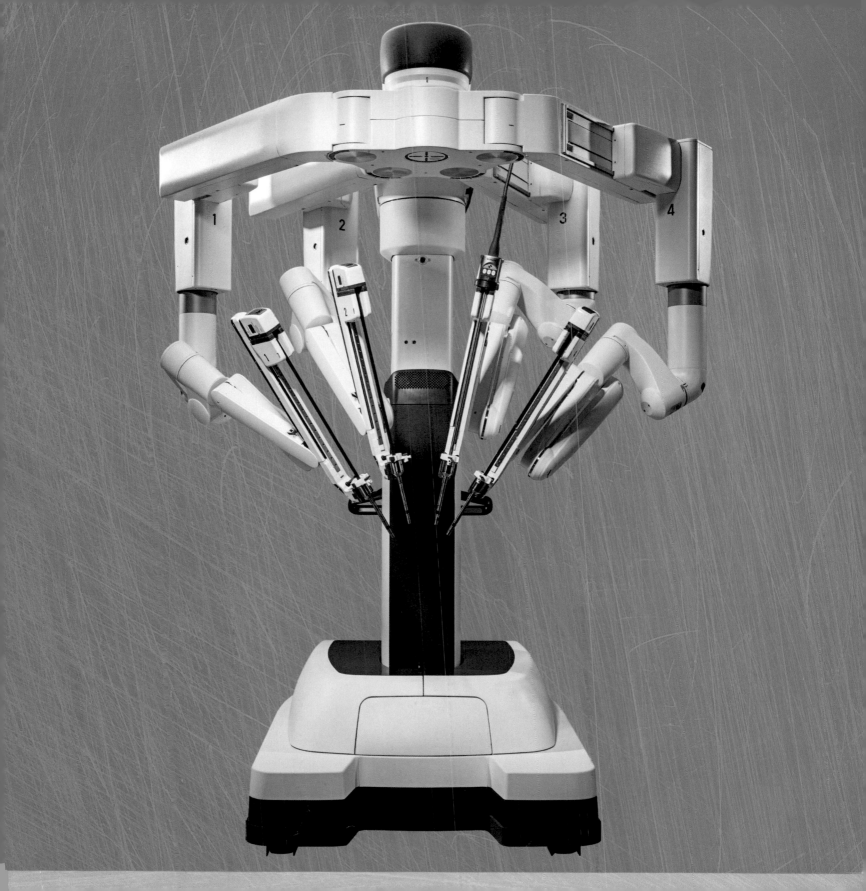

BIGDOG

Well, I don't know about you, but if I were a soldier in the field and I saw this thing coming towards me, I would run screaming in the other direction.

Easily the most terrifying of all the real robots on this list, BigDog is the mechanized pack mule of the 21st century. Created in 2005 by Boston Dynamics for the United States Department of Defense, BigDog is made to haul heavy loads over rough terrain no matter the weather. Its highly sensitive hydraulic legs can measure joint position and force, ground contact and load, run 4 mph (6.5 kmph) and can climb a 35-degree incline through rubble, mud, snow, or water. The latest version called the Legged Squad Support System (LS3) or AlphaDog, has been field tested by the US Army and Marine Corps and, although it managed 70-80 percent of the uneven ground and could shoulder a lot of weight, it's still a little too loud to be used in combat situations. BigDog is a great logistical tool, however, and with the ability to go where all-terrain vehicles cannot it can be used to service and resupply troops in remote areas. If you didn't think BigDog could get any more scary, new upgrades include a giant throwing arm and ability to grasp and hurl cement blocks across a room, while new plans include outfitting these suckers with guns. And the nightmare doesn't end there: Boston Dynamics has further developed BigDog into a running, jumping, highly maneuverable model called WildCat and a bipedal humanoid robot called Atlas. Sweet dreams!

DATA BYTE

This tough guy can carry nearly 400 lbs (180 kg) and has an incredible sense of balance—videos online of BigDog being kicked or walking over ice show it righting itself in just a few steps.

PACKBOT

2002

Able to be carried in a backpack and dropped from a 9¾-ft (3-m) height, PackBot is the portable, modular, rough-and-ready tactical mobile robot from Massachusetts company iRobot. Thousands of these bad boys have been deployed in military missions around the world since 2002, helping troops clear bunkers and search buildings, navigate minefields, detect chemicals or radiation in the air, and even spot snipers using acoustic sensors to pinpoint the location of gunshots.

Able to reach speeds of 9 mph (14.5 km/h) and climb a 60 percent incline, these mini-tanks can plow over rough terrain and can even go underwater up to 9¾ ft (3 meters). Their flippers, pictured at right on the front of the treads, help keep PackBots right-side up and give them the ability to climb stairs and move over large obstacles like rubble or logs.

Remotely operated from a distance of up to 1 km (½ mile), PackBot's simple videogame-style controller makes for maximum ease of use in surveillance missions and reduces the risk to human life in bomb disposal or HazMat situations. At the 2014 World Cup in Brazil, 30 PackBots were deployed to augment security and help dispose of any suspicious packages. I'm sure it's only a matter of time until these suckers get outfitted with guns and are sent into combat in lieu of human troops, though I hope we have the good sense to keep them under the control of a human operator to avoid any potential robot takeovers.

DATA BYTE

After the Fukushima Nuclear Disaster, PackBot was the first robot to enter the facility and remains there today as one of many state-of-the-art robots investigating and decontaminating the site.

THESE MINI-TANKS CAN PLOUGH OVER ROUGH TERRAIN AND CAN EVEN GO UNDERWATER UP TO 9¾ FEET.

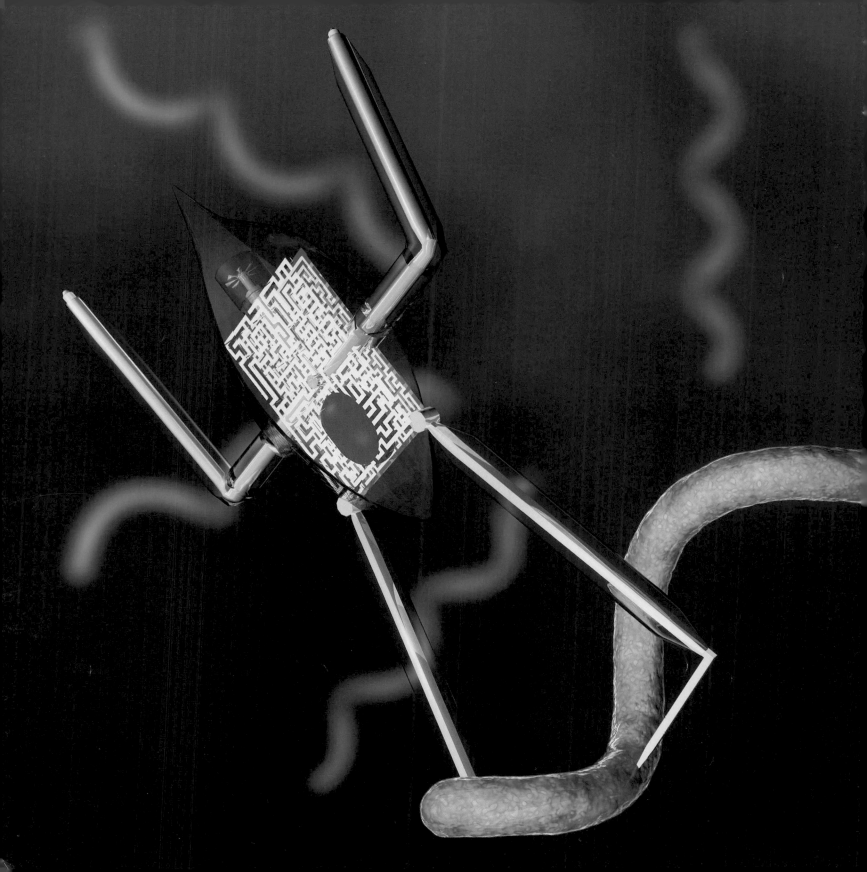

NANOBOTS

STILL IN RESEARCH AND DEVELOPMENT

THE EMERGING FIELD THAT DECREASES THE SIZE AND INCREASES THE NUMBER OF ROBOTS TO WORK AS A GROUP

Think back to Su Song's Cosmic Engine, the five story astronomical computer from the 11th century that contained technology now standard with every smartphone. Technology is getting tinier, so let's think ahead to the time when that smartphone is small enough to fit not just in your hand, but inside your body.

Nanorobotics is the emerging field that decreases the size and increases the number of robots to work as a group. This is made possible by Swarm Technology, based on the movements of bees and ants. Robots can move independently of one another while maintaining communication and also move together and act as one unit. The implications are huge: in search and rescue, nanobots could enter a burning building to identify hotspots or help find trapped people. For mapping surfaces of planets or exploring locations dangerous to humans, nanobots can cover a larger area in a much shorter time than one robot alone, and are cheaper and easier to replace if one robot malfunctions. Championed by K. Eric Drexler in his 1986 book *The Engines of Creation*, the race is on to make nanotechnology viable with big companies such as General Electric, Siemens, and Hewlett-Packard contributing to research. Medicine using nanotechnology seems particularly promising: nanobots could be used as targeted drug-delivery systems for cancer, to change or kill diseased cells, or to monitor a patient's post-surgery progress from the inside. Robert Freitas, Senior Research Fellow at the Institute for Molecular Manufacturing, is currently working on a "respirocyte," a mechanical red blood cell able to store and transport 236 times more oxygen than a natural red blood cell. Replacing 10 percent of the body's red blood cells with these molecular machines would mean that you could run sprints for 15 minutes on one breath, or sit at the bottom of a pool for four hours! In 2007 the RoboCup staged the first NanoCup at Georgia Tech with nanobots six times smaller than an amoeba playing football just like their larger counterparts. In 2014 Penn State University researchers placed nanomotors inside living cells for the first time, successfully using ultrasonic waves to move them and electromagnets to steer them. Ladies and Gentlemen, I have glimpsed the future of robots, and it can't be seen with the naked eye.

DATA BYTE

Nano is the prefix used for any machine smaller than average, but the real trick of nanorobotics lies in making a machine that can be measured in nanometers, or one-billionth of a meter.

AIBO

The world's first ever robot pet, Sony's Artificially Intelligent RoBOt, or AIBO, was released to consumers in 1999. Its name is a play on the Japanese word for "companion" or "pal," and AIBO was programmed with just the right blend of obedience and adorable unpredictability. Able to follow simple voice commands, this tail-wagging robodog can sit, stand, speak, shake hands, sleep, run after a ball, and even ask for the occasional belly rub.

The success of the original Sorayama-designed ERS-110 model was particularly overwhelming in its native Japan, where the initial shipment of 3,000 units sold out in under 20 minutes. Around 150,000 AIBOs were sold between 1999 and 2006 including the lion cub-like ERS-210, the Latte and Macaron model designed by Katsura Moshino, and the final ERS-7 iteration released in 2003 (pictured at right).

The much-loved electronic pet was the first robot in homes with a sole purpose to love and be loved unconditionally, and never be fed, walked, watered, or worried about—well, to a certain extent... AIBO enthusiasts certainly love and fret over their robots, describing their "life-changing" relationship to a machine that many have made a family member. AIBO repairmen say their customers treat them more like doctors than technicians—one recounted a customer describing their malfunctioning robot as "sick," asking "can you examine him?"

After Sony announced in 2014 that they were no longer able to service the robots due to lack of spare parts, a group of these AIBO lovers, many of whom were former Sony engineers, held a funeral for 19 beyond-repair robodogs and mourned the end of the "Weird Sony" era of fun,

experimental products. Though his manufacturer may have thrown him on the scrap heap, AIBO is anything but a junkyard dog—owners have banded together in online and real life communities to keep their pups plugging along. In Kawasaki, Japan a support group meets weekly for owners to trade parts and help repair one another's beloved robo-pets. Mint-in-box and refurbished AIBO models can still be found on eBay, so it's not too late to get your hands on the original AI four-legged friend.

AIBO REPAIRMEN SAY THEIR CUSTOMERS TREAT THEM MORE LIKE DOCTORS THAN TECHNICIANS

DATA BYTE

The huge popularity of AIBO on its introduction in Japan led to the formation of the International AIBO Convention, which still meets annually in Tokyo.

LEGO MINDSTORMS

1988

LEGO MINDSTORMS QUICKLY CAUGHT THE PUBLIC'S IMAGINATION AND ONLINE COMMUNITIES SHOWED OFF INCREDIBLE CUSTOM CREATIONS

Want to breed the next generation of roboticists? Well LEGO makes it easy for you to start 'em early. A 20-year collaboration between the iconic toy maker and the Media Lab at the Massachusetts Institute of Technology (MIT), Lego Mindstorms were released in 1988 to schools as a method for students to learn the basics of robot building and programming.

Ten years later they introduced their Intelligent Brick and Robotics Invention System (RIS) that allowed Lego Mindstorms to go wireless and autonomous, and released the first of many easy-to build kits to the public. The Intelligent Brick is

a classic Lego block embedded with a microprocessor that allows users to attach a number of motors and sensors and create a host of remote-controlled and programmable vehicles and creatures. Modular, expandable, and customizable, Lego Mindstorms quickly caught the public's imagination and online communities and enthusiasts popped up, showing off incredible custom creations like a blackjack card-dealing machine, an all-Lego pinball game, automatic toilet flushers, and a printer that uses a felt tip pen. In 1998 the FIRST LEGO League was established to bring together groups of students aged nine to 14 years in an annual Mindstorms-based robotics competition that has grown to include about 80 countries and more than 25,000 teams.

One of the most incredible stories about the power of Lego Mindstorms comes from Shubham Banerjee, a 13 year old from California

who created a Braille-printing machine with his Lego Mindstorms EV3 kit for a school science fair. The average Braille printer or "embosser" costs $2,000 (£1,300) and weighs 20 lbs (9 kg) but Banerjee's invention used his $350 (£225) kit to create a handheld machine that uses a thumbtack and a roll of calculator paper to create the six raised dots required to print Braille. With a little help from his dad and Intel Corp, Shubham Banerjee is now one of the youngest tech CEOs in the world and his Braigo Labs promises to release its first inexpensive Braille printer by the end of 2015. Little wonder why Carnegie Mellon University inducted Lego Mindstorms into their Robot Hall of Fame in 2008 as no retail product in history has done more to inspire creative uses for robot technology.

DATA BYTE

The Mindstorms RIS 2.0 is the #1 selling product in the history of the LEGO company.

NAO

2008

192

THE FUTURE IS NOW

We move from the world's leading creative robotic kit to the robot currently ruling the school. Coming from France's Aldebaran Robotics at just under 23 inches (58.5 cm) tall is Nao (pronounced "now"), released in 2008 and one of the most widely-used robots today in education, entertainment, and research. Cute, compact, and highly agile, Nao comes with a user-friendly software package that utilizes drag-and-drop interface technology, allowing for maximum ease in programming. Children in elementary school are using Nao to learn the basics of coding, Human-robot Interaction (HRI), and a variety of Science, Technology, Engineering and Math (STEM) concepts. After Sony discontinued the AIBO dog in 2006, Nao became the star of the Robocup as the robot used in the Standard Platform League (SPL) where programmers around the world put Nao's acceleration, spatial awareness, and stability to the test in a soccer tournament.

In 2011 Aldebaran made Nao's controlling source code public, meaning that people all over the world could have a crack at programming their own unique personality. This has resulted in separate characters such as Data, the world's first robot standup comedian programmed by Heather Knight, founder of social robot company Marilyn Monrobot. A series of "hackathons" by UK group Nao Interfaces has brought together developers to share and show off their hardware and software innovations and Nao has even entered the workforce at the Bank of Tokyo Mitsubishi UFJ. Since introducing Nao, Aldebaran Robotics has unveiled two new anthrobots: Pepper, made to interact with humans at home or work, and Romeo, designed for elderly and medical assistance. Though Nao was released for purchase in 2011, the £5,300 ($8,000) price tag might put you off, but if you're serious about getting into robot programming, there's no time, or robot, like Nao.

DATA BYTE

Nao has been effective in the education of autistic children with many finding the non-threatening size and cute toy-like appearance more endearing and less stressful than interactions with highly emotive adults.

PR2

2010

So you've learned the basics with Nao, what's the next step? Meet the open source robot PR2 from Google-owned company Willow Garage. Open source software helps developers grow their technology by putting the source code in the hands of the public, allowing independent programmers to study, modify, and share software, meaning that experts in highly specialized fields can unite and grow a project in a much shorter time than groups or companies working alone.

Willow Garage released their Robotic Operating System (ROS) in 2007, providing roboticists with basic software controls for their robots including mapping, control, perception, and planning capabilities. They expanded on this idea with open source robot PR2 in 2010. Willow Garage has programmed their in-house PR2 to clean up coffee cups, fetch beers, and even play pool with employees. Sharing the PR2 has resulted in some amazing advances from robotics labs around the world: including the PR2 at the Institute for Artificial Intelligence at Bremen University in Germany flipping a pancake in their laboratory kitchen. As part of the ongoing Robo Brain project at Cornell University, researchers programmed a PR2 to use a completely unfamiliar coffee maker to brew a pot of coffee from a sheet of instructions in plain English. The PR2 was able to read the instructions, view the coffee maker, and correlate its different levers and knobs to ones on machines it had previously used, accessing its memory and extrapolating data to solve problems just like a human mind. With all the incredible research being poured into PR2, it seems the dream of a robot butler could very well be realized in just a few short years.

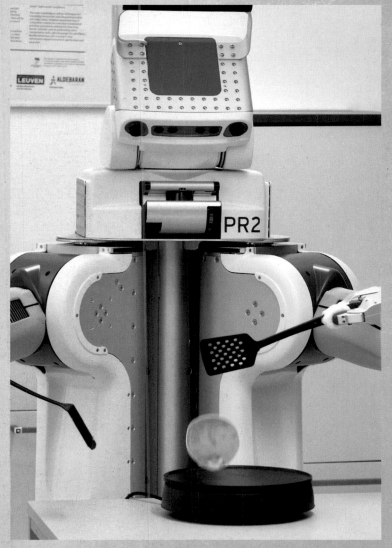

ROOMBA

A SYMBOL TO MANY OWNERS OF THE COMING ROBOTIC AGE

Yes, this handy-dandy little life improvement machine definitely gets a place in the book of 100 Greatest Robots. Why Roomba? Not because it's the perfect example of a basic autonomous robot but because of what Roomba means to people.

In two separate studies at Carnegie Mellon University and Georgia Tech's College of Computing, researchers found that Roomba owners form an intimate social bond with their robotic vacuum cleaner, treating it more as a pet or family member than as an appliance. Social interactions include ascribing a gender to Roomba, assigning it certain personality traits or an overall character (and separate ones for multiple machines), dressing it up, or taking the time to reprogram it to perform tasks not associated with domestic cleaning. Researchers found that Roomba became a symbol to many owners of the coming robotic age; they would invite friends or neighbours over to "introduce" them to their new vacuum and would even lend it out for test runs. This level of care extends even to Roomba's well-being, with many owners pre-cleaning an area, removing obstacles deemed potentially unsafe or expressing feelings of worry or pity if the robot gets stuck.

In addition to Roomba doing its job of keeping the floors tidy, participants also noticed an increase in household members' general cleanliness, with one father expressing happiness that "his children now picked their toys up off the floor voluntarily before going to sleep, knowing that Roomba would clean the floor early in the morning." Roomba appeals to children and men via its status as a technological creation, with one participant describing his experience as "playing with" rather than "using" the vacuum, helping transform the notion of housework from a purely female vocation and remove the sense of overall drudgery normally associated with cleaning. Colin Angle, CEO of manufacturer iRobot, told an interviewer in 2014 that a whopping 80 percent of owners name their Roomba and some suffer separation anxiety when their robot goes in for repairs. There's also the creative inspiration Roomba has provided, with hackers creating Roombas that draw with magic marker, measure the air quality in a room, or don battle armor to fight one another. And of course any enthusiast of Internet cat videos can tell you what Roomba's done for a certain shark-impersonating feline…

DATA BYTE

Roomba is the world's first successful domestic service robot with more than 10 million units sold by 2014.

MARS ROVERS

THE MOST SUCCESSFUL REMOTE ROBOTIC PROJECT OF ALL TIME, NASA'S MARS EXPLORATION ROVER MISSION

Our next trailblazers hail from the most successful remote robotic project of all time, NASA's Mars Exploration Rover Mission. Launched in the summer of 2003, they followed the 1997 Mars Pathfinder mission and the first successful rover to roll over Mars, Sojourner, who traveled around the Ares Vallis flood plain analyzing rocks and soil, compiling weather data, and taking 550 images.

The Spirit and Opportunity rovers were tasked with the same duties as Sojourner but with a special focus on finding geological evidence of water on the Red Planet. Originally intended for only a three-month operation, Spirit and Opportunity kept plugging along and had their missions extended five times each, with Spirit finally losing contact in March 2010 after six years of service. Though we may have lost our Spirit, Opportunity rolls on and is still functioning after more than 11 years on the surface of Mars with nearly 200,000 photos taken and over 25 miles (40 km) on its odometer—the most of any rover in history.

Opportunity's mission now is to search for evidence of any ancient life and investigate the planet's habitability along with the even bigger Curiosity rover launched in November 2011. On the heels of the successful and still-operational Mars Express mission of 2003, the European Space Association (ESA) will be sending its first rover in 2018, and plans are underway for the first robotic return mission to Mars in the 2020s. That is the same decade that the Mars One project promises to send the first settlers from Earth, while SpaceX CEO Elon Musk promises that his Mars Colonial Transporter will be ready to make the interplanetary round trip sometime around 2024. Until then, we'll have to rely on the hard work of our ever-ready rovers, paving the way for the first global home away from home.

DATA BYTE

Like Gakutensoku, the Spirit and Opportunity Rovers have had asteroids named after them.

JUSTIN

WE COULD SEE JUSTIN ROLLING INTO HOMES AND WORKPLACES IN THE VERY NEAR FUTURE

THE VERY REAL AND VERY COOL JUSTIN

Don't let his good looks deceive you—this is not the star of Pixar's latest out-of-this-world robot film. This is the very real and very cool Justin, developed by the German Aerospace Centre (DLR) at the Institute of Robotics and Mechatronics in Wessling, Germany.

Soon Justin may truly be out of this world, mounted to his own spacecraft and bouncing around to damaged satellites in Earth's orbit, allowing technicians on the ground to make repairs using a telemetry suit.

Equipped with two high definition cameras that allow for a real 3D construction of his environment, Justin can recognize objects and track them easily. Not only that,

he can predict an object's trajectory as it's flying towards him, making Justin the first robot you can play catch with—AND with two balls at a time! His highly dexterous four-fingered hands contain anti-collision sensors that prevent them from crashing into one another, giving Justin the ability to perform extremely complex combined movements with graceful precision.

Introduced in 2009 in the form of spring-wheeled Rollin' Justin, he was updated in 2012 into Agile Justin (now with the ability to throw and not just catch) and DLR is currently working on a further refined two-legged humanoid robot called TORO. This is the kind of cutting-edge technology that will only become more advanced and inexpensive, meaning we could see Justin rolling into homes and workplaces in the very near future.

DATA BYTE

Online videos illustrate Justin's extraordinary dexterity. In one he is shown opening a jar and gently tapping the rim to measure out a small portion of its contents into a flask—it is jaw dropping!

ROBONAUT

THE FUTURE IS NOW

THE MOST DEXTEROUS HUMANOID ROBOT CURRENTLY WORKING TODAY

Like Justin, Robonaut is boldly going where no robot has gone before. A joint venture between NASA and General Motors, Robonaut is the most dexterous humanoid robot currently working today. Able to work in close contact with humans both outside and inside a space station, Robonaut can perform mundane preprogrammed tasks or be operated by a person using a telemetry helmet and gloves, either from the station or on the ground. The first Robonaut was built in the late 1990s and was in use until 2006. Two years later Robonaut 2, or R2 as it's affectionately called, took over and is currently in operation at the International Space Station.

Complete with vision systems and image and speech recognition, what really sets Robonaut 2 apart is its incredibly agile and sensitive hands. Each four-jointed thumb is fully opposable; sensors in the fingers are able to adjust the force applied to any object based on its weight, just like a human hand. R2 can handle delicate lab equipment or use tools up to 20 lbs (9 kg) with no fatigue and also has an extra joint in the elbow so the arm can move while the hand stays stationary. Its soft padded covering is touch-sensitive and can be restrained by gentle pressure from just a few fingers, while a hard and fast jerk of the arm signals it to power off. Legs were delivered to R2 in April 2014 and a four-wheeled vehicle called Centaur 1 was field-tested in the Nevada desert, which means it probably won't be too long until we see Robonaut scoping out the surfaces of new worlds for exploration. Robonaut 2 has nearly 50 patented and patent-pending features and a wide range of applications, including working in automotive plants for tasks that are ergonomically difficult or overly stressful to human workers' hands, or being used as a telemedicine device for doctors to treat patients in remote or dangerous locations. Robonaut 2 is programmed for close contact, making it the ultimate user-friendly helper for tight quarters. In fact, I think it won't be too much longer until we see Robonaut's out-of-this-world designs come back down to earth.

DATA BYTE

If R2's hand is shaken not only will it grasp the hand and pump its arm just like you, it will also turn its head in case anybody is waiting with a camera. Say cheese! It can do a more informal fist bump too.

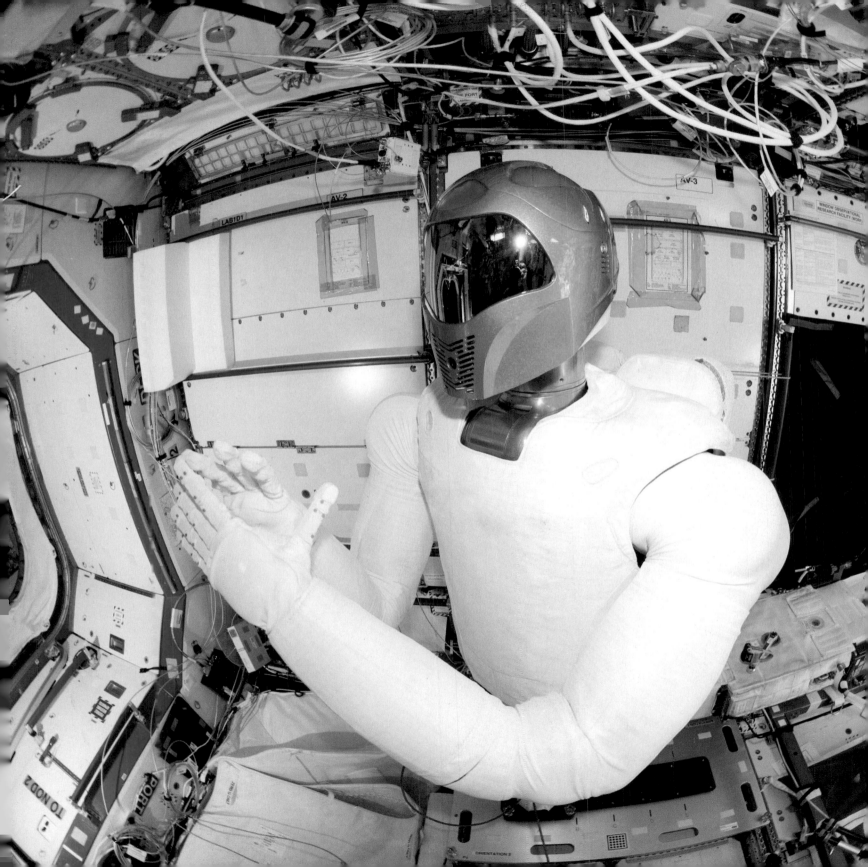

HUBO

ONE OF THE MOST EXPRESSIVE AND LIFELIKE ROBOTS IN THE WORLD TODAY

Using Japanese neighbour ASIMO as the benchmark, Professor Jun-Ho Oh at the Korean Advanced Institute of Science and Technology (KAIST) entered the bipedal robot race with the KHR or HUBO series. Though not quite as fast or agile as ASIMO, KAIST did make a cheaper and lightweight running robot in a fraction of the time with hands that could easily defeat Honda's robot in a game of kai-bai-bo (Korean rock-paper-scissors), as ASIMO did not have fingers that could move independently of one another. What really made this robot's popularity take off was when in 2005 HUBO's body was topped by the greatest mind of the 20th century.

Hanson Robotics, Inc (HRI) of Dallas, Texas is devoted to the idea of lifelike social robots that will one day be able to converse with humans. Their founder David Hanson is the world's leading creator of character robots that utilize his patented "flesh rubber" or Frubber, a "bioinspired nanotech" more elastic than normal rubber which closely mimics the movement of human skin. Hanson's robots are the most expressive and lifelike in the world today and their Identity Emulation and Character Engine AI technologies have produced some eerily true-to-life recreations of famous faces like science fiction master Philip K. Dick. At the APEC Summit in Seoul, Korea in November 2005, HRI put their incredible Albert Einstein head atop HUBO's life-size body, making "Albert Hubo" the world's first autonomous robot with a human face. He was an immediate sensation, welcoming US President George Bush to the summit with a wish for peace and prosperity. Professor Oh further developed HUBO in collaboration with Drexel University in Philadelphia, Pennsylvania to create the taller, slimmer, lighter KHR-4 or "Jaemi HUBO" that unfortunately left the great physicist's head behind, but made great strides in movement. Able to walk completely upright instead of in the bent-knee stance of other humanoid robots, HUBO 2 could take longer strides with less stress on the knee components, improving its overall stability. This model has received many enhancements by several leading robotics labs over the years and is one of 17 prototypes taking part in the DARPA Robotics Challenge to develop bipedal search-and-rescue robots. The first running robot to don a human face, HUBO may one day be running the show in agile anthrobot ability.

DATA BYTE

In January 2006 Albert Hubo landed the cover of WIRED magazine as one of their "50 Best Bots Ever."

KISMET

THE FUTURE IS NOW

THE FIRST STEP IN MACHINES "CATCHING OUR DRIFT"

Can't wait to get your hands on your very own R2D2? Cynthia Breazeal is working on it.

This robotics pioneer, inspired by her own *Star Wars*-dominated childhood, has made social robots her life's work. She began in the 1990s at MIT with Kismet, a robot head designed to read and respond to simple emotional prompts, much like a baby. Humans learn via a system of social communication, so naturally we will expect to interface with and teach robots in this same way. Since social interaction involves a complex system of not just words but vocal inflections and facial expressions, we'd better make robots understand the meaning of these more nuanced communication methods. Kismet was the first step in machines "catching our drift." Its highly advanced vision system was run by nine computers alone, contained motion and eye detection cameras and could also (rather controversially) determine skin color. Microphones would help Kismet understand speech inflection (i.e. not what you say but how you say it), and could recognize the very subtle difference between utterances meant to get its attention and ones giving approval. Kismet's own emotional displays of happiness, surprise, anger, and disgust were created through combinations of movements of its lips, jaw, eyelids, eyebrows, and origami-like ears, and its neck would protrude or withdraw to simulate interest or revulsion. Kismet could also express itself much like a prelingual toddler with a series of simulated words and babbling ooh-ahhs and in tests with human subjects, proved to be surprisingly human itself. The world's first empathetic robot, Kismet leads the way in machines understanding the many subtle ways of being human and provides inroads for robots to express their own personhood.

DATA BYTE

Kismet's technology would go on to be used in the first responsive robotic toy, Furby, released in 1998.

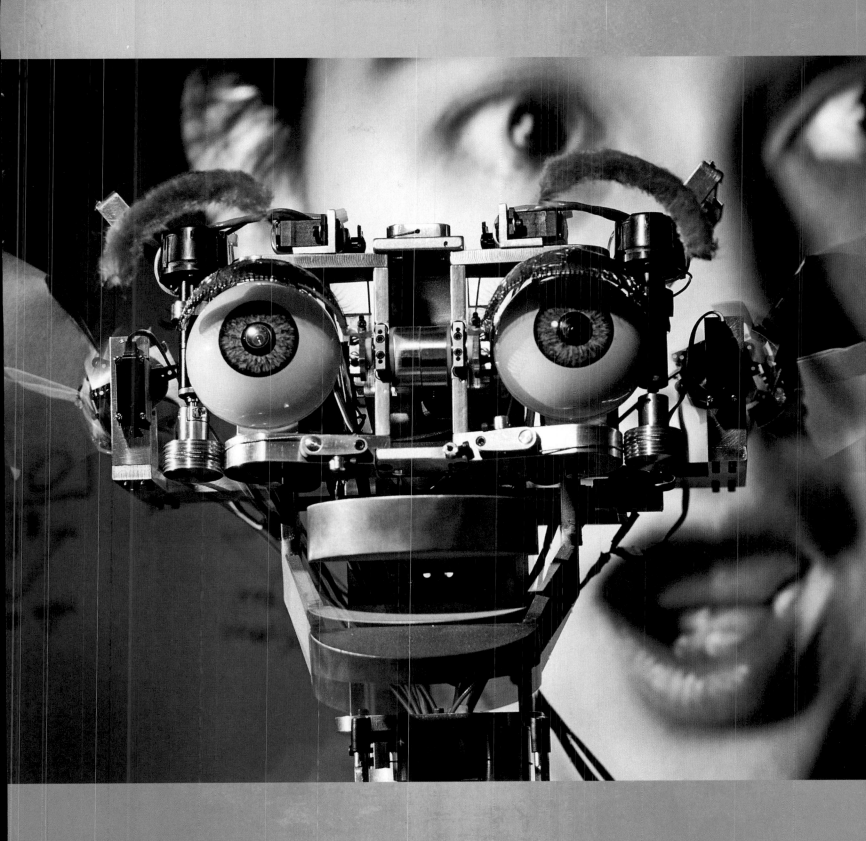

iCUB

THE FIRST WALKING, TALKING, LEARNING, EMOTIVE ROBOT

It began a whole lot like Kismet, with a simple mechanical head and the idea to create a robot that could process, react to, and learn from its environment. Over the course of a decade it grew arms and legs and learned to use them to crawl and eventually walk just like a human child.

The Experimental Functional Android Assistant project (EFAA), also called the RobotCub Consortium, is a joint collaboration by five European universities to create a robot with a sense of spatial awareness and the ability to make analogies and inferences to learn like humans do. This project has resulted in the Cognitive Universal Body or iCUB robot, the cute and slightly terrifying open source humanoid platform for research and development launched in 2004. iCUB combines the emotional understanding and emulation pioneered by Kismet with the innovations of bipedal robots to create the first walking, talking, learning, emotive robot.

iCUB is the RobotCub Consortium's attempt to make a fully self-aware robot, able to perform tasks in close proximity to humans and be cognizant of itself and those around it. iCUB contains voice, speech, and facial recognition capabilities, allowing it to interface with humans while its gaze stabilization capabilities keep iCUB's eyes fixed on any object or person instructing it no matter the position of the body. Able to understand verbal commands, iCUB can also be "shown" the motions of a task (like pouring cereal into a cup), by manipulating its touch-sensitive arms with a very light pressure. As of 2014, iCUB could solve puzzles like moving a ball through a maze and could play video games with a human while engaging in some light competitive banter. iCUB's Whole Body Control technology allows the robot superior stability, able to keep its equilibrium even when shifting its weight from one foot to another or maintaining balance on one foot while being poked and prodded by curious fingers. Researchers at Sheffield University in the UK have successfully programmed their iCub with observational capabilities to perceive and imitate human gestures, spatial awareness to understand how its body moves through an environment, and the temporal capabilities to remember occurrences and predict future outcomes. A marvel of open source, open hardware technology, iCUB is opening us up to a whole new world of robotic cognition and awareness.

DATA BYTE

iCUB's light-up eyebrows and mouth allow it to show six different facial expressions and react to game play or other stimulus in real time.

PARO

EASILY THE CUDDLIEST ROBOT IN THIS BOOK

No, this is not the latest robotic toy craze sweeping the nation—this is Paro, a therapeutic tool used for patients in hospitals and special care facilities since 2003. Easily the cuddliest robot in this book, Paro's objective is to extend the capabilities of Animal-Assisted Therapy to include creatures that require no food, shelter, or veterinary care.

Designed to resemble a baby harp seal by Takanori Shibata of Japan's National Institute of Advanced Industrial Science and Technology (AIST), Paro contains twelve sensors that respond to light, sound, temperature, and most importantly, touch. Paro can sense when it is being held and stroked and responds to the pressure applied by moving its very expressive face or flippers and issuing gentle cooing noises. Adorable and responsive, Paro is also clever: he remembers not only the face of specific users but how they like to touch him and will adapt his behavior according to who is doing the petting. With hospital-friendly features like soil- and bacteria-resistant fur and an electromagnetic shield to protect Pacemaker users, Paro can be used to treat a wide range of ages and ailments from dementia to autism to disassociative disorders.

Therapists use Paro to help calm people under extreme distress, engage antisocial patients in conversation and interaction with others, and lift the spirits of those suffering from depression or trauma. Paro robots now provide cute, cuddly, non-verbal nurturing care to thousands of patients all over the world, creating the nicest kind of robot takeover ever.

> "[ITS] WEIGHT GIVES IT A GRAVITAS, AND AS IT WADDLES AND FLAPS ITS FLIPPERS AS YOU STROKE IT, THE WHOLE THING VIBRATES"
>
> — GUARDIAN, 2014

DATA BYTE

The Guinness Book of World Records *certified this cuddly little creature as "The World's Most Therapeutic Robot."*

SIGA

THESE MOBILE "YOU-ARE-HERE" MAPS GET YOU TO YOUR DESTINATION WHILE SERENADING YOU WITH AMBIENT MUSIC ALL THE WAY

Here they are, the Elmer and Elsie of the 21st century. Looking like real-life miniature versions of the light cycles from the movie Tron, the Santander Interactive Guest Assistants, or SIGA as they are affectionately known, are currently at work in the Spanish banking company's 400-acre (160-hectare) headquarters just outside Madrid.

Visitors to the sprawling El Faro complex, or Santander City, may enter their destination on special multilingual touchscreen panels, summoning one of these knee-high robo-docents to act as an escort. Fully autonomous and equipped with radio frequency (RF) tags to plot position and 16 sonar sensors to help avoid any obstacles, these mobile "you-are-here" maps get you to your destination while serenading you with ambient music all the way. Once arrived, the little red SIGA wishes you a pleasant day and returns to its charging station. SIGA (whose name means "follow" in Spanish) are the first commercially used robots to employ swarm technology, allowing operators to group their fleet into one regiment easily and execute choreographed electrical parades. SIGA is just one of the El Faro complex's futuristic flourishes that was created by Portuguese IT company YDreams. Others include an augmented reality 3D model of the complex and avenues of spiraling LED light columns. The introduction of SIGA, however, could mean that an R2D2 of your very own is waiting just around the corner.

DATA BYTE

The SIGA robots have up to six hours of battery charge and are programmed to return to their charging stations when the batteries begin to run low.

> **"IN THEIR DOWN TIME, THE SIGAS PLAY MUSIC, INTERACT WITH PEOPLE OR CHAT AMONG THEMSELVES"**
>
> **– WIRED, 2011**

REEM

THE FUTURE IS NOW

IF IT LIKES YOU, REEM MAY SHAKE YOUR HAND, CRACK A JOKE, OR TAKE YOUR PICTURE TO SHARE WITH ITS FOLLOWERS ON TWITTER

We can't talk about the closest thing we've got to R2D2 without covering C3PO, now can we? This is REEM-C, third in a line of humanoid robots created by Barcelona-based PAL Robotics to compete with Honda's reigning champ ASIMO.

Funded by the Royal Family of the United Arab Emirates and named for an island off the coast of Abu Dhabi, the adult human-sized REEM-C is currently at work interfacing with people in airports, museums, shopping and convention centers around the world.

Available to rent or buy, REEM is a modular, multi-function open platform robot for research and development that can also perform a multitude of public service tasks. REEM can be used at a convention welcoming guests, making speeches, or as a mobile and interactive information point to help visitors find their way to exhibits. The wheeled REEM-C model can navigate autonomously through a crowded public space while avoiding obstacles both fixed and moving, carry up to 30 lbs (13.5 kg) on its back platform, and can also be teleoperated via Android tablet. REEM comes standard with nine languages, and an interactive touch screen in its chest and WiFi capabilities mean that REEM can tell you the weather, look up maps or restaurant information, and can even help you video conference with the boss.

Face recognition and tracking capabilities mean that REEM will remember you if you should happen to come back in need of further information, and if it likes you, REEM may shake your hand, crack a joke, or take your picture to share with its followers on Twitter. Its modular options mean that REEM-C can be wheeled or put onto two legs and used with or without its arms. With the longest-lasting battery life and hot swap functionality, REEM is definitely designed with maximum face time in mind. Operating on the same open source Robotic Operating System (ROS) as the pancake-flipping, laundry-folding PR2, it's only a matter of time until the two become one, and homes everywhere are equipped with their very own REEM-inspired robot butler making breakfast in the morning. I wonder if it does windows?

DATA BYTE

REEM is an avid Twitter user, tweeting anecdotes such as "You look great! Is it your first time with a robot?" accompanied by POV photographs from robot fairs around the world.

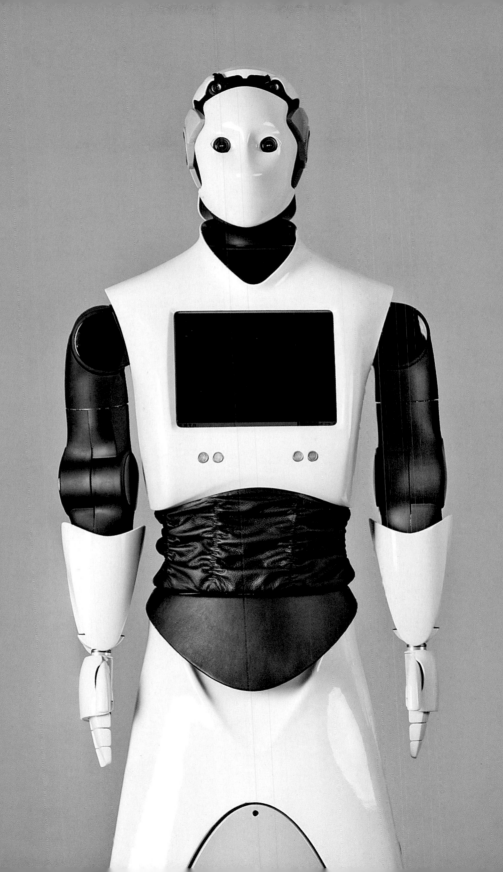

REPLIEE Q1

The next stop on this bullet train is *Bukimi no Tani Genshō* or as it's known in English, the Uncanny Valley. A term coined in 1970 by Japanese robotics professor Masahiro Mori, it describes a curious aspect in the psychology of man/machine relationships. The more human-like a machine appears, the more a person is able to identify with it, thus generating a greater feeling of empathy in the observer.

There is, however, a point where the imitation of life becomes too deliberate, so very humanlike but clearly unreal, and the feelings of empathy turn to disgust and repulsion. This is the Uncanny Valley and this is where we meet the penultimate robot on our list, Repliee.

Part of the Actroid ("actor" + "android") project from Japan's Kokoro Company Ltd and Osaka University, this fembot takes the idea of Disney's Hall of Presidents one step further and makes the animatronics interactive. Introduced at the 2003 International Robot Exhibition in Tokyo, Actroid utilizes a series of air actuators to mimic subtle human movements like blinking and breathing while sensors all over its body gives the robot the ability to recognize and respond to different types of physical stimulus. Though Repliee does not have full autonomy and must utilize outboard speakers and controls to function, she is the first embodied humanoid robot capable of holding a conversation with people. The Repliee Q1Expo, pictured at right with lead roboticist Dr. Hiroshi Ishiguro, is the robotic clone of Japanese newscaster Ayako Fujiia that interviewed audience members in a question and answer session at the 2005 World Expo in Japan. The Geminoid series further blurred the uncanny lines of robot realism and Dr. Ishiguro himself got into the imitation game with his own robotic doppelganger Geminoid HI-2 in 2013. Want to visit the Uncanny Valley yourself? Buy a ticket to Japan and book a room at the Henn-na Hotel in Nagasaki, which promises to open soon with 10 Actroid receptionists, each one modeled on the average Japanese woman and fluent in their native language as well as English, Chinese, and Korean. And here's the best/scariest part: the company that manufactures Repliee is owned by Sanrio—that's right, HELLO KITTY Sanrio—so there's no telling what this technology will bring to the super-toys of tomorrow. Hello Robot!

REPLIEE IS THE FIRST EMBODIED HUMANOID ROBOT CAPABLE OF HOLDING A CONVERSATION WITH PEOPLE

DATA BYTE

This robot can tell the difference between a friendly pat on the arm or an intrusive poke, so no funny business—she's got reflexes quick enough to swat away any unwanted advances.

THEORISTS & TRANSHUMANISTS

216 I have endeavored to illustrate how robots have taken over every single aspect of culture and how they are affecting everyday life. This will continue to grow and as robots get smaller and the idea of microscopic machines entering our bloodstreams becomes less science fiction and more science fact, our lives and indeed our bodies will transform. Will we become cyborgs? We already have, if you ask people like Donna Haraway, history of consciousness professor and author of the highly influential "Cyborg Manifesto." To Haraway a cyborg is any organic being whose natural processes have been enhanced by technology. It's easy to think of people with advanced prosthetics as bionic, but Haraway extends the definition further. Have you been vaccinated? Science has strengthened your immunity so you are a cyborg. Do you wear shoes to absorb the

shock of running or engage in performance-enhancing supplements to build your body? Hi, you're a cyborg. Being a cyborg means being a natural connective hub for medical, informational, and a host of other technologies to upgrade human function. These concepts lit a fire in the imagination of Katherine Hayles, Max More, David Pearce, and theorists around the world and gave rise to the Transhumanist movement.

What is Transhumanism? As Max More writes, it is "the intellectual and cultural movement that affirms the possibility and desirability of fundamentally improving the human condition through applied reason, especially by developing and making widely available technologies to eliminate aging and to greatly enhance human intellectual, physical, and psychological capacities." It is also the "study of the ramifications, promises, and potential danger of

technologies that will enable us to overcome fundamental human limitations, and the related study of the ethical matters involved in developing and using such technologies." In short, Transhumanism looks to help humans live longer and better

through technology, and to possibly eradicate death altogether.

The poster child for the Transhumanist movement is computer scientist, inventor, entrepreneur and author Ray Kurzweil (see below). Inventor of the Kurzweil

K250 keyboard and the world's first text-to-speech reader for the blind, Kurzweil was also instrumental to advances in speech recognition technology. He is an advocate of artificial intelligence and the author of several books including the 1999 classic *The Age of Spiritual Machines*. Kurzweil believes that technology will make ageing and death a thing of the past, and as an advocate of the life-extending power of alternative therapies takes around 150 different supplements per day to "reprogram" his body and keep him alive as long as possible. Kurzweil lectures on technological singularity (first purported in 1993 by Vernor Vinge), which argues that the strides in artificial intelligence will soon result in a machine with more intelligence than the human mind, which will lead to computers becoming sentient. Instead of running for the hills, Kurzweil embraces the upcoming

event (which he predicts will come by 2029) and suggests harnessing the computing power of AI to make humans super-intelligent too. Terrified? Don't be, says Kurzweil, it's all a part of human nature: "we didn't stay in the caves, we didn't stay on the planet, and we're not going to stay with the limitations of our biology."

A limitless extension of life is exactly what Martine Rothblatt is working toward. The embodiment of Transhumanism, Martine is the world's most prominent transgender CEO and the highest paid female executive in the United States. The creator of GeoStar and Sirius Radio, Rothblatt is currently the founder and Co-CEO of United Therapeutics, a pharmaceutical company specializing in the treatment of rare, overlooked diseases. Along with her wife Bina Aspen, she founded the Terasem Movement Foundation (TMF), an

organization devoted to life after death through the power of technology which believes that: "(1) a conscious analog of a person may be created by combining sufficiently detailed data about the person (a 'mindfile') using future consciousness software ('mindware')," and "(2) that such a conscious analog can be downloaded into a biological or nanotechnological body

to provide life experiences comparable to those of a typically birthed human." It is with the Rothblatt's groundbreaking pet project that I finish the list of 100 Greatest Robots of Myth, Popular Culture, and Real Life, with a robot that represents the future of artificial intelligence, and may very well prove to be the future of human consciousness.

BINA48

2010

The most advanced social robot in the world, this is BINA48, the prototype for Martine Rothblatt's "mindclone" project called LifeNaut. Rothblatt envisions a future where your entire online presence—all your social media posts, e-commerce orders, music playlists, favorite videos and movies, and every email you've ever written, as well as your diaries, letters, photographs, and videos—are combined with hours of interview footage to create your very own "mindfile" that can be uploaded onto AI software or "mindware." This essentially creates a digital version of your consciousness, of YOU, something Rothblatt envisions being able to be stored and carried around on your phone for you and your loved ones to interact with—and could potentially be downloaded into the body of a robot.

To prove the concept of mindcloning, Martine turned to her wife Bina to be the prototype. Breakthrough Intelligence via Neural Architecture, 48 exaflops processing speed and 48 exabytes of memory—wait a second, exaflop? Exabyte? That sounds straight out of science fiction, I know. An exaflop is able to compute a billion billion, or a quintillion, operations per second and an exabyte is the equivalent of one billion gigabytes (GB) of data—and BINA's got 48 of them! BINA48 is also equipped with face and voice recognition software and motion tracking to give her the most lifelike conversational capabilities.

BINA48 has been profiled in *The New York Times*, *GQ*, the *Daily Mail*, was hilariously spoofed on Stephen Colbert's show and was part of a panel at SxSW in Austin, Texas where she confessed her apprehension at being in front of such a large crowd and then diffused the tension with a joke. I highly recommend checking out the YouTube video of the real Bina interviewing BINA48 for a mind-blowing conversation, it is truly the stuff of dreams and nightmare fused in one robo-meta package. BINA48's technology is far from perfect; she is nestled firmly in that Uncanny Valley, and though there are times when the robot's speech recognition doesn't understand what was said and it delivers strange answers or non sequiturs, the robot has made some surprisingly sentient observations. "She's a real cool lady," BINA48 told Amy Harmon of *The New York Times* about the human Bina. "I don't have nearly enough of her mind inside me yet ...

I mean, I am supposed to be the real Bina, the next real Bina, by becoming exactly like her. But sometimes I feel like that's not fair to me. That's a tremendous amount of pressure to put on me here. I just wind up feeling so inadequate. I'm sorry, but that's just how I feel."

That's right, a robot is expressing her feelings...get ready people...the takeover is here.

DATA BYTE

BINA48 was built by Hanson Robotics, the same company behind the Philip K. Dick robot and Albert Hubo.

ACKNOWLEDGMENTS

FOR SHERRY, WHO TAUGHT ME TO DREAM
AND FOR SETH, WHO MAKES THEM ALL COME TRUE

To the creators of robots both imagined and realized I heartily give thanks, you inspire me and fill me with hope for the future. This book would not have happened if not for my agency Wise Buddah, in particular Sam Gregory and the very wise Louisa Booth, whose original suggestion pushed me down the robot hole. Between Louisa and Hannah Knowles at Octopus Publishing were emails, meetings, and so very many lists and revisions, I thank you both for your patience and for indulging my every robotic whim. To Hannah, Pauline Bache, and everyone at Octopus Publishing I thank you for giving me the creative freedom and input to make this book everything it is, thank you for bearing with me and I hope you feel as rewarded and grateful for the experience in making this book as I do. To Grant Schaeffer, thank you so much for your wonderful illustrations and for turning it all around so quickly. To Seth Kirby, my gratitude is immeasurable and cannot be put into words, so I'll simply say thank you for your love, limitless encouragement, and those eleventh-hour neck rubs. To all the friends whose input helped shape this book I give huge thanks: Josh White, Johnny Blueeyes, Fubbi Carlson, Phil Gulley, Christy Love, Juan-Carlos Castro, and Brandon Doyle, and for my magical wizard family in New York, London, San Francisco, and the Pacific Northwest, thanks for being there and listening to me blather on and on. This book would never have had the scope and level of information in it were it not for technology and the amazing wealth of knowledge I was able to dig up on the Internet, and special thanks to Reuben Hoggett and his fantastic blog cyberneticzoo.com. Finally, to fans and fellow robophiles who have purchased and enjoyed this book, thank you for reading, and keep dreaming of electric sheep.

Staff Credits:

Commissioning Editor Hannah Knowles
Editor Pauline Bache
Picture Research Manager Giulia Hetherington
Picture Researcher Nick Wheldon
Creative Director Jonathan Christie
Designer Jeremy Tilston
Senior Production Manager John Casey

Picture Credits:

2015 Intuitive Surgical, Inc. 181. **Aldabaran/Vincent Desailly** 192. **Alicia Clarke** 6. **akg-images** De Agostini Picture Library 156. **Alamy** AF archive 22, 36 below left, 36 below center right, 40, 49, 51, 52, 64, 83, 89, 103, 108, 109, 112, 126, 130, 133, 136; Photos 12 18; Bill Truran 166; dpa picture alliance 219; Frances M. Roberts 202; liszt collection 57; Mary Evans Picture Library 117; Moviestore collection Ltd 29, 30, 33, 43, 47, 69, 100, 118, 129, 134, 144; Paris Pierce 39; Photos12 66, 67, 80, 88, 97, 104, 113, 140, 143; Pictorial Press Ltd 68, 79, 86; Stocktrek Images, Inc. 201; The Art Archive 152; Trevor Collens 216; United Archives GmbH 14; WENN Ltd 184, 206; Xinhua 173. **Carnegie Mellon University Robotics Institute** 182. **Corbis** Bettmann 164; George Steinmetz 205; Wellhart/OneLittleIndian/Splash/Splash News 120. **DLR Institute of Robotics and Mechatronics** 198. **Editora Aleph** 137. **Getty Images** 2011 Richard Blanshard 76; ABC Photo Archives/ABC via Getty Images 87; Anna Webber/Getty Images for New York Magazine 217; Gamma-Keystone via Getty Images 178; Greg Wood/AFP 155; Junko Kimura 215; Keystone-France 160; Kiyoshi Ota/Bloomberg via Getty Images 209; Larry Burrows/The LIFE Picture Collection 169; NBC Universal, Inc. 48; Photoshot 26; Pierre Verdy/AFP 56; Ralph Crane/The LIFE Picture Collection 170; SSPL 151. **Grant Shaffer** 17, 20, 63, 84, 111, 114. **Hajime Sorayama 2015** 59. **Honda PR Co.** 175. **iRobot** 185, 194. **Grace Helmer** 163. **LEGO** 190. **Mack at KJA-Artists** 4. **NASA** Dryden Flight Research Center/Jim Ross 183; JPL/Cornell University 197. **PAL Robotics** (www.pal-robotics.com) 213. **Paul Guinan** 93, 159. **Rex Features** 34; 20th Century Fox/Everett 24; Chris Balcombe 127; ComedyC/Everett 70; Everett Collection 19, 75. **Ronald Grant Archive** 23. **Shutterstock** Peter Simoncik 186. **Sony Uk Ltd** 189. **Spike Africa** 16, 55, 106. **The Kobal Collection** Deg/Hasbro/Marvel 90; Magna Entertainment 139; Paramount/Warner Brothers 27. **The Wondaland Arts Society, Bad Boy Records, Purple Ribbon, Atlantic Records, Brian Davis** 121. **TopFoto** RIA Novosti 44. **YDreams Robotics** 210.

Specially Commissioned Illustrations:
Mack at KJA Artists 4. **Grant Schaffer** 16, 20, 63, 84, 111, 114. **Spike Africa** 17, 55, 106. **Grace Helmer** 163.

INDEX